BECOMING A BETTER LEADER AND GETTING PROMOTED IN TODAY'S ARMY

Books by Wilson L. Walker

Becoming a Better Leader and Getting Promoted in Today's Army

The Complete Guide to the NCO-ER

Self-Development Test Study Guide

Up or Out: How to Get Promoted As the Army Draws Down

BECOMING A BETTER LEADER AND GETTING PROMOTED IN TODAY'S ARMY
The NCO's Guide to Putting the Soldier First

Wilson L. Walker
Master Sergeant, U. S. Army, Retired

IMPACT PUBLICATIONS
Manassas Park, VA

BECOMING A BETTER LEADER AND GETTING PROMOTED IN TODAY'S ARMY
The NCO's Guide to Putting the Soldier First

Copyright © 1997 by Wilson L. Walker

All rights reserved. Printed in the United States of America. No part of this book may be used or reproduced in any manner whatsoever without written permission of the publisher: Impact Publications, 9104-N Manassas Drive, Manassas Park, VA 20111-2366, Tel. 703/361-7300.

Library of Congress Cataloging-in-Publication Data

Walker, Wilson L.
    Becoming a better leader and getting promoted in today's army : the NCO's guide to putting the soldier first / Wilson L. Walker
    p. cm.
    Includes bibliographical references and index
    ISBN (invalid) 0-942710-69-2 (alk. paper)
    1. United States. Army—Non-commissioned officers's handbooks. 2. Command of troops. 3. Leadership   I. Title
U123.W35 1997
355.3'38'0973--dc21   96-46964
                                                                  CIP

For information on distribution or quantity discount rates, Tel. 703/361-7300, Fax 703/335-9486, or write to: Sales Department, Impact Publications, 9104-N Manassas Drive, Manassas Park, VA 20111-2366. Distributed to the trade by National Book Network, 4720 Boston Way, Suite A, Lanham, MD 20706. Tel. 301/459-8696 or 800/462-6420.

CONTENTS

Preface vii

Acknowledgments viii

PART I

BECOMING A BETTER LEADER

CHAPTER ONE
The Making of a Leader 1
- The Noncommissioned Officer 1
- Know Yourself First 3
- Know Your Values 5
- Know Your Beliefs 8
- The Meaning of Leadership 10
- The Factor of Leadership 10
- Leadership Competencies 12
- What the Leader Must BE, KNOW and DO 15
- Leadership Styles 20
- Understanding Leaders Relationship 22

CHAPTER TWO
Knowing and Training Your Soldiers 23
- Putting Your Soldiers First 25
- Influencing Your Soldiers 26
- Getting To Know Your Soldiers 30
- Soldiers Values 32
- Ingredient For a Combat-Ready Team 34
- Building the Combat-Ready Team 36
- Training the Combat-Ready Team 41

iii

- Training the Team During Combat 45
- Sustaining the Combat Ready Team 47
- Sustaining the Team During Combat 48

CHAPTER THREE
Team Training With Higher Headquarters 53
- Principles of Training 53
- Mission Essential Tasks List (METL) Development 55
- Planning for the Mission 57
- Short-Range Planning Briefing 59
- Near-Term Planning 60
- Execution and Assessment 61
- Evaluation of Training 62

CHAPTER FOUR
Who We Must Fight and Why 64
- New Concepts in Army Doctrine 64
- The Army Objectives 65
- Constitution 65
- National Security 66
- United States Policies on NBC 66

PART II
GETTING PROMOTED

CHAPTER FIVE
Junior NCO's Promotion (SGT and SSG) 71
- Semicentralized Promotion 71
- Key Events and Work Scheduling 71
- The Promotion Packet 73
- Becoming Eligible for Promotion 73
- Recommendation for Promotion 73
- Eligibility Criteria 74
- Time in Grade and Time in Service 74
- Rules for Conducting the Promotion Board 75
- The Study Guide 76
- The Women's Class A Uniform 78
- The Men's Class A Uniform 79
- Awards in Order of Precedence 81
- More Facts About Junior NCO's Promotion 82

CONTENTS

CHAPTER SIX
Senior NCO's Promotion (SFC, MSG, SGM) 84
- Centralized Promotion 84
- Eligibility Criteria 85
- Centralized Promotion Board 86
- Personal Appearance and Written Communication 86
- Board Results 87
- Monthly Promotions 87
- Service Obligation 88
- Reclassification Prior to Promotion 88
- Preboard Process for NCO's in Zone of Consideration 89
- Standby Advisory Board 89
- DA Photos and Military Records 92
- NCO-ER 94
- More Facts About Senior NCO's Promotion 94

CHAPTER SEVEN
Removal and Reductions 96
- Removing a Soldier From the Standby Promotion List 96
- Removal From a Centralized Promotion List 97
- Processing Command Initiated Removals 98
- HQDA Initiated Removals From a
  Centralized Promotion List 99
- Reduction in Grade 100
- Reduction for Misconduct 101
- Reduction for Inefficiency 101
- Conducting a Reduction Board 102
- Soldiers Rights to Appeal 103

CHAPTER EIGHT
NCO Education System (NCOES) 105
- Reason for the NCOES 105
- Fitness Before Training 106
- Reading Before Leading 106
- Requirement for PLDC Attendance 107
- Course Instruction for PLDC 107
- Requirement for BNCOC Attendance 108
- Course Instruction for BNCOC 108
- Requirement for ANCOC Attendance 109
- Course Instruction for ANCOC 109
- Requirement for FSC Attendance (not part of NCOES) ... 110

- Course Instruction for FSC 114
- Requirement for SGM Academy Attendance 115
- Course Instruction for the SGM Academy 116

Appendix A **119**
- Nonpromotable Status 119
- Delay of Promotion due to DA Form 268 121
- NCO's Security Clearance Requirements 122
- Most Usable Terms 122

Appendix B **126**
- Abbreviations 126

Index **129**

About the Author **132**

PREFACE

To become a better leader in today's Army, you need to master the art of taking care of your soldiers.

The only reason you get promoted, a good assignment, or have an outstanding NCO-ER or OER or are selected for one of the leadership schools is because someone feels you will be able to care for your soldiers.

When Barry Sanders was selected as the Associated Press Offensive Player of the Year in 1994, he made the statement *"I was just part of the group. There are a lot of guys on the team who made me look good."*

Every successful leader in the Army became successful because other leaders were aware of his dedication to care for his soldiers, and in return, his soldiers made him look good. The leaders who are not promoted to the next grade are passed over because other leaders do not feel they are taking care of their soldiers. The key to becoming a better leader is taking care of your soldiers by guiding them to accomplish the tasks and missions which will make them better soldiers. The reward of being a good leader will show in your troops. This book will help you get started on how to become a better leader.

Thoughout this book we use the word "he" to refer to male and female leaders and soliders. Please note that no sexism is implied nor intended. We fully recognize there are many successful leaders and soldiers in the Army, regardless of gender. Our decision to use the masculine pronoun was prompted by stylistics, not chauvinism.

ACKNOWLEDGMENTS

After thanking the one above us all, my Mother, wife, children and Tom Hasselstrom who put it all together, I must thank the following for their up-to-date information.

CSM Tony M. Daniels
Commandant, III, Corps NCO Academy

CSM Charles E. Hermon
Command Sergeant Major, 4th Battalion, 43rd ADA

1SG John L. Driver
Asst. Commandant, III, Corps NCO Academy

1SG William L. Peterson III
First Sergeant, HHB, 4th Battalion, 43rd ADA

Mrs. Maureen D. Vice
Education Counselor, Fort Hood

PART I

BECOMING A BETTER LEADER

1

THE MAKING OF A LEADER

THE NONCOMMISSIONED OFFICER

It is said that the Noncommissioned Officer (NCO) is the backbone of the Army, but, what is a Noncommissioned Officer and why do some NCOs say another is not a Noncommissioned Officer but rather a Sergeant? Is it the same as saying he is a daddy, not a father? When one NCO says another NCO is not a Noncommissioned Officer, but a Sergeant, it is his way of saying the NCO has poor leadership skills. There are two classifications of leaders in the Army: Commissioned Officers and Noncommissioned Officers.

The Commissioned Officer, ranking from lieutenant to general, is appointed by the President of the United States to act as his legal agent and help carry out his duties as the Commander-in-Chief. They receive their legal authority from the President. The Army classifies a Noncommissioned Officer as one who holds rank from corporal to sergeant major. The Noncommissioned Officer's authority is delegated to them by the Commissioned Officer, which means the NCO serves as an agent for the Commissioned Officer.

Authority is the legitimate power of leaders to direct subordinates or to take action within the scope of their responsibility. NCOs also have command and general military authority. Each mission demands for Commissioned Officers and Noncommissioned Officers to work together. The Commissioned Officers must give the NCOs guidance, resources, assistance and supervision necessary for them to do their duties. When working together, the Commissioned Officer and Noncommissioned Officer learn from each other. Years ago, when I was a platoon sergeant in Germany, our platoon received a new platoon leader who was a young female lieutenant.

I'm the type of leader who wanted to run my outfit my way. I had the best platoon in the unit and possibly the Battalion. I did everything I could to keep my distance from my lieutenant, because she was always asking questions regarding the platoon or would always want me to handle tasks that were of little importance to the mission. When she would come down to the section, I would always try to get her to leave, for one reason or the other.

The Commander called me into his office to ask me what the problem was between me and my platoon leader. After I spoke to him, he explained to me that I had something that most NCOs wished they had—a leader that wanted to learn how to run an outfit.

After that session with the Commander, I spent many hours showing my platoon leader the ins and outs of running a platoon. I would have the rooms set up so that she could inspect them and even asked her to put on a pair of coveralls, so that she could do PMCS on one of the platoon trucks.

One day, around lunch time, the General flew onto our site unexpectedly and my lieutenant was the only officer present. She looked at me and I said, "Ma'am, I suggest you report to him, give him an assessment of the mission, and escort him wherever you want him to go." She escorted the General, gave him a quick briefing of the mission, and was given a good report for the visit. That not only made me feel good, but also the Commander and all of the other officers were confident of her good performance.

When I was in the process of leaving that duty station, I was given a "going away" party by some of the NCOs and my platoon leader was present. She stood up and stated that I was a very good Noncommissioned Officer, and that I had helped her become a better leader. I can't begin to tell you how good that made me feel. Officers receive additional training by NCOs; if you get a weak leader, it is possible he was trained by a weak NCO.

The NCOs are responsible for assisting and advising officers in carrying out their duties. They advise Commanders on individual soldier's proficiency and training requirements needed to ensure unit readiness. Because of this, Commanders are free to plan, make decisions and program future training and operations. The officer plans and decides what needs to be done—the NCOs make sure it gets done.

Another type of officer in the Army is the Warrant Officer. Although they are appointed by the Secretary of the Army, rather than by the President, they have the same authority, duties and responsibilities as Commissioned officers and are eligible to hold command positions.

The major role of the Noncommissioned Officer is the training and care of the soldiers. How they care for the soldier has changed somewhat, due to the huge advances in technology and quality of life programs. The NCOs of today must be well-educated so they can understand, speak and execute the directives of the officers. They must also enforce the standards set forth in regulations and directives, and be a positive role model for others to follow.

Noncommissioned Officers are now being placed in a combat role with more challenges than before. When they find themselves the only leader present on the battlefield, they must be able to execute the intent of the officer leader in order to achieve success. They must be bold, willing to take risks in decision making and be selfless and totally professional. Most important of all, the Noncommissioned Officer must be able to take the initiative to execute and accomplish the assigned mission effectively in the absence of the Commissioned Officer. The Army doctrine demands that the Noncommissioned Officer be proficient in all leadership areas.

KNOW YOURSELF FIRST

To know yourself is to know your values, beliefs and ethics. It is also being totally honest and giving yourself a 100 percent true evaluation on your actions and abilities. This is something only you can do.

If you truly know yourself, you will be able to form a relationship with your boss and soldiers that can make you the successful leader you want to become. You can be a leader that can train soldiers for war and accomplish the unit, Army and national objectives.

Your superior wants a person of leadership skills who:

- Will have order within his formation during his absence
- Can be trusted to get the job done
- Can care for his soldiers
- Will instill the values and beliefs similar to his own

A soldier wants a leader who:

- He can look up to
- He can respect and who will respect him
- Will reward him when he meets or exceeds the standards of the mission
- Will discipline him when he is wrong
- Will stand up for him when he needs the support
- Will train him so he is able to accomplish any given mission, and, at the same time, accomplish his personal objectives

Most of all, soldiers want a leader that is competent, skilled, proficient and able to make sound decisions under stress. As you can see, your soldiers only want what you are supposed to give them, and, it may very well be the same things you expect from your leaders. The more you know and understand about yourself, the better your chances are of becoming a great leader. Are you the type of leader you would like to be? Ask yourself the following questions:

- Am I happy with my job performance?
- Am I giving 100 percent to the mission?
- Do my soldiers respect me?
- Am I a competent leader?
- Are the standards I set achievable?
- Am I a leader or a follower?
- What is my number one priority?
- Can my leader and soldiers rely on me?
- Do I listen or do most of the talking?
- Do I use "micro-management" to get things done?
- Am I an optimist or a pessimist?
- Am I a selfless leader?
- Do I take the credit for my soldiers deeds?
- Do my peers see me as a sergeant or an NCO?
- Do I have a good relationship with my boss?
- Do I have a good relationship with my soldiers?
- Do I make decisions just to please my leaders?

THE MAKING OF A LEADER   5

- Do I lead by example?
- Am I a "do as I say, not as I do" leader?
- Do I feel the unit can't make it without me?
- Am I truthful to my leader and soldiers?
- When does the day end for me?
- Do I run my soldier away when he wants to talk to me?
- Do I have the last say?
- Do I skip out of training?
- Is my military bearing in accordance with Army standards?
- Am I aware of what I must BE, KNOW and DO?
- Do I over or under supervise my soldiers?
- Am I technically and tactically proficient?
- What is my leadership style?
- What is my superior's style of leadership?
- Is the QMP a threat to me?
- Did I earn my last promotion?
- Did I earn my awards?
- Am I accountable for my soldier's actions?
- Do I work along with my soldiers?
- Do I know how all of my soldiers are living?
- Can I out score most of my soldiers on the PT test?
- Do I understand the factor of leadership?
- Do I understand my unit mission?
- Is my training realistic?
- Do I push my soldiers towards a promotion?
- Do I train my soldiers as a team?
- Do I read at at least a 12th grade level?
- How important are my soldiers to me?
- Am I a role model to my soldiers?

As a leader, it is important that you know yourself so that you can maximize your strengths and work to improve your weaknesses. The old saying, "No two people are alike", also holds true for leaders. You must know what you are capable of doing, as a leader, if you are going to be successful in training your soldiers as a team to accomplish the mission.

KNOW YOUR VALUES

Values are our personal, private and individual beliefs about what is most important to us. If a promotion is the most important goal you want to achieve, then most likely you will get promoted, because you will do all you can to get promoted.

The same thing holds true about your soldier's care, your military bearing or maybe the unit mission. Values govern your entire life-style from how you dress, the car you drive, your leadership style, the food you eat and even how your raise your children or lead your soldiers. Your values will also influence your priorities and decisions.

The four values that all soldiers are expected to possess are courage, candor, competence and commitment. **Courage** is standing up for what you feel is right or wrong; however, your feelings do not necessarily make the issue right or wrong. Will you take a stand or will you think how a situation may affect your career and look the other way? I remember a time when I was an acting First Sergeant at a direct support unit in Germany and encountered a situation requiring the use of courage.

The Commander was never around, and, when he was there, he was just that....there! I first approached the Commander about the lack of support he was portraying to the unit and, when all conversations were ignored, I called the Battalion Commander and told him that I needed to speak to him with the Captain present. At the same time, I told the Captain that he and I had to see the Battalion Commander.

We reported to the Lieutenant Colonel's office, I told him how things were being handled at the unit and that he had a Captain there, but not a commander. The LTC asked me to return to the unit and he spoke with the Captain for another two hours.

After that day, conditions improved at the unit and the Captain carried no resentment towards me for addressing the issue. If he did, he certainly didn't show it! Physical courage is overcoming fears of bodily harm and doing your duty. Overcoming fears can be greatly reduced with realistic, exciting and meaningful training.

Candor is being honest, open and sincere with others. It is the ability to use tact to get your point of view across and, at the same time, accept the decisions of your leaders, as long as they are legal and proper orders; even if you do not personally agree with them.

Competence is proficiency in required professional knowledge, judgment and skills. If you do not know your job, believe me when I say your soldiers will be the first people to know it. If you do not know the things your job requires for you to know, by all means, do not hesitate to seek assistance from someone or break out the manuals that will help you learn your job. Being assigned to another job will not help, because sooner or later, the lack of knowledge will catch up to you.

Commitment is carrying out all unit missions, but remaining loyal to the unit, Army and National duties, are all part of the professional Army ethics. The leader with the most commitment is the one that is the most successful. It is the quality of commitment which separates the good from the best.

A soldier that was in a good unit will always talk about the good things the unit leadership did, and that adds up to the fact that the unit had more commitment. The best NCO or officer in your unit is the one with the most commitment. Why do you think they are always tasked for the most important missions? If they weren't capable of doing the job, they would have been passed over in receiving the task.

Most people receive their values from the environment around them as they were growing up. From the time you were born, your parents taught you the same values that were instilled into them. Most of our values were based on the punishment-reward technique. Parents carrying strong education values reward their children with money or other things for good grades they bring home from school. However, if the grades were not good, the child was grounded until an improvement was noticed. The child may not have done something wrong; his values are simply not the same as his parents. There is a similarity in the Army.

A Noncommissioned Officer who is an expert with his weapon may want all of his soldiers to become experts. If he can attain the maximum score on the PT test, he'll push for all of his soldiers to maximize theirs.

It is important that you know your soldier's values because they may or may not be the same as your own. If they are not the same as yours, that doesn't mean they are wrong. As your goals or self-image changes, so will your values. If your goal is to become a sergeant major and you succeed in this goal, you will have more control over many NCOs and may expect more from them than you did before becoming a sergeant major. There are some people who spend their time with a different class of people and may change the way they walk or talk to match their "new" self-image. Many of those values are unconscious. You will create this new image because you feel you have to.

People generally feel uncomfortable of individuals who have values much different from their own. Your most important value will always be first. If the value of a friendship exceeds honesty, you may lie for a friend. If honesty is your most important value, however, you will tell the truth at the possible expense of friendship.

It is important to know your values so you will be able to motivate, support and direct yourself at the deepest level. A soldier that values life and freedom may do all he can to keep from going into a combat zone.

I was told of one NCO who went to Desert Storm, jumped from the top of a van and broke his leg, so he could return home. There is no better way to bond people, than to align them through their highest values. Discovering your soldiers values is simply a matter of finding out what's important to them. A team with winning values will win, every time.

KNOW YOUR BELIEFS

Beliefs are assumptions or convictions you hold as true regarding circumstances, concepts or a person. Beliefs are the compass and map that guide you toward your goals and give you the security of accomplishing them. Nothing is more powerful in directing the force in human behavior than a belief. To change your behavior, you have to begin with your belief. This statement also holds true for your soldiers.

For example, there were two staff sergeants in my platoon in West Germany who would show up late for work. I spoke with them regarding the importance of being on time, but the problem continued. I wrote up a counseling statement stating that if this problem was not resolved, I was going to submit paperwork recommending a summarized Article 15. The two staff sergeants received a summarized Article 15 because they **believed**, that because of their rank, their continued lateness would be overlooked.

I instructed both of them to report to the unit at 0400 hours each morning, on time, for one week. As it turned out, one of the staff sergeants had to sign in on Christmas morning. After the punishment was imposed, the two staff sergeants never reported late to work again, during my tenure as platoon sergeant. They changed their behavior because, now, they had the belief that if they didn't come to work on time, it would affect their career, and they were right.

Beliefs are like commanders of the brain, and the brain simply does what it's told. If you tell yourself that you can or can't do something, you're right. Maybe you know someone in your unit that keeps falling out of the morning PT run. If you watch them closely, you will soon find that they will stop at just about the same place every time they do so. The reason is that they **believe** that they just can't go any farther.

If someone runs with them and encourages them along the way, they will find that they can go farther or complete the run, because now, the other person is in control of their mind, and their belief has now been over-ridden. When they are on their own again, however, many times they will go back to stopping at the same place, because they have it in their mind that they can't make it.

If you want them to make the run, encourage them to run about another 100 yards. Notice where they stop, and add the hundred yards. Tell them, "Don't worry about completing the run. All I want you to do is run up to that parking lot (or whatever spot it may be) and back to the unit." Many times, they will now make the run, because now they will believe that they only have to go another hundred yards. They will change their belief and their mind will tell the body to keep going up to the parking lot. After a week, or so, add more distance, and, before you know it, that soldier will be making the run with the unit.

If you don't think your soldiers program themselves as to how far to run, the next time that you have them out on the morning run, run them a few blocks and then run them past the point you normally stop running. If you look back, you will find that some of your soldiers have stopped running.

Beliefs will do one of two things; support you or limit you. If you say that you think you can't do something, you choose the beliefs that limit you. If you say or think that you can do something, you choose those beliefs that support you. It is your beliefs that determine how much of your potential you'll be able to tap. Whenever you say "I can't", you give your brain a command not to do something, and your failure is pre-determined.

Beliefs are like values, in that they start with your environment. If all you see is failure and despair, it's very hard to change your beliefs to become successful. If you work on it a step at a time, however, you will become the leader you want to be.

You can create the beliefs that will allow you to be successful. Historians inspire present day soldiers with their renditions of the Battle of Gettysburg in the leadership manual, and units have boards like Audie Murphy and Sergeant York. The best way to create a belief in yourself is to do something once. That's why the soldier that attains a maximum score on the PT test will do it over and over again, without practicing.

To be a successful leader, you must choose your beliefs carefully. If you are convinced you are going to fail, you will not put your full efforts into succeeding. The first step toward becoming a good leader is to find the beliefs that will guide you into becoming one. The road to success consists of knowing the end results of your efforts, taking action, measuring your progress and having the flexibility to make changes until you're successful.

THE MEANING OF LEADERSHIP

The manual on Army Leadership (FM 22-100), dated June 1973, states that military leadership is the process of influencing men in such a manner as to accomplish the mission. The same manual, dated October 1983, states that military leadership is the process by which a soldier influences others to accomplish the mission. Each manual states that leadership is the process of influencing others, be it men, women or soldiers.

The latest leadership manual, dated July 1990, states that military leadership is the process of influencing others to accomplish the mission by providing purpose, direction and motivation. As you can see, the latest manual goes a step farther by saying, "influence others by providing purpose, direction and motivation." Let's explore why.

Purpose gives the soldier a reason why he should do difficult things under dangerous and stressful circumstances. Direction outlines tasks to be accomplished, based on the priorities set by someone in a leadership position. Motivation gives the soldier the will to live up to his capabilities and to accomplish the mission.

At boards to determine the Soldier of the Month, Soldier of the Year, or eligibility for promotion, the question is often asked, "Which is more important, the soldier or the mission?" The answer is the soldier. The reason for this is because no mission can be accomplished without the soldier, and the leadership manual states that leadership is the process of influencing others to accomplish the mission, not accomplishing the mission by influencing others. The mission, however, is the end product of influencing the soldier, and is the number one goal. Later, I will further discuss why your soldiers must be your number one priority.

THE FACTORS OF LEADERSHIP

The factors of leadership consists of the "led", which are your soldiers; the "leader", which is yourself; the "situation", which is the duty or mission; and communication. These four factors of leadership are always present, and will affect the actions you take.

Those who are **led** are the first factor of leadership. They (soldiers) cannot all be led the same way. Despite all their military indoctrination and training, they are still individuals. You must take your leadership style into account, as well as their beliefs, values and your ability to influence them. You must know each of your soldiers' competence, motivation and commitment, so you can take the proper actions. You must influence, persuade and encourage them to actively participate and help accomplish the mission. This is much easier if they respect, trust and have confidence in you, as their leader.

The **leader** is the second major leadership factor. In order to be an effective leader, you must know yourself, your values, beliefs, job, soldiers and leadership style. You must know how to motivate your solders. You must know what you must KNOW, BE and DO.

Leadership is a twenty-four hour a day, seven day a week obligation. You must know your ups and downs, your strengths and weaknesses and your abilities and limitations so that you can have complete control over yourself, first, and then your soldiers, so you have a better chance of getting the mission accomplished.

The **situation** is the third major leadership factor. There will be many situations you, as a leader, will have to deal with, from a soldier not showing up for duty, to trying to get your leaders to go along with your plans. You, and only you, will have to determine the best actions to take, and will have to shoulder the responsibilities of their success or failure. Again, this is a matter of knowing yourself and your soldiers.

You must be skilled in identifying and assessing the situation, so that you can take the right actions at the right time. Remember, it takes actions to produce results, and nothing can be done without some type of action being taken.

The fourth major leadership factor is **communication**. Your level of communication will determine your level of influence on your soldier. The way you communicate with your soldiers is important because, during the first few minutes, they will zero in on what they see—your skin color, gender, age, appearance and your facial expressions. Your eye contact, body movement, the personal space you take, and touch will also be noticed.

When you communicate, nearly half of what you are communicating is done through body language and facial expressions. After that, the next thing your soldiers will be attentive to is your voice; the rate or tempo, loudness, pitch, clarity and tone. Your actual words are generally what is paid the **least** attention. All the time you are communicating with them, they will be gathering bits and pieces of information about you, to form a complete picture. Any time a leader violates the soldiers expectations, he assumes risks.

Skin color still remains the dominant factor of appearance. In situations where you think your skin color may be a negative factor, seek to counter the stereotype by paying extra attention to your appearance, facial expression and eye contact.

Gender also fosters stereotypes—males are perceived to carry more power, authority and credibility in first encounters. Female leaders can express their authority by being prompt, having a strong handshake, maintaining direct eye contact and using a smile to counter any excessive aggressiveness.

Age is neither positive or negative and depends entirely on the soldiers' expectations of the leader.

Appearance is much more than the way you dress. It can also take into account your body type, posture, hair, accessories you wear or carry, smell and the appearance of your footwear.

Facial expressions are the cue most soldiers use to pick up on the leader's mood and personality. It's important that your facial expression is matched with the tone of your voice and the words you are speaking. Make eye contact, but adjust it to your comfort level, and be aware of your movements. Leaders who are sure of themselves have active, persistent movements. They look strong and confident, take up more personal space and freely move into the space of others.

The consequences of touching are often risky and are subject to a lot of different interpretations. A salute or firm handshake is, then, the safest and least controversial. Self-assessment, study and experience will improve your understanding of the four factors of leadership.

LEADERSHIP COMPETENCIES

The nine leadership competencies provide a framework for leadership development and assessment. They are areas where the leader must be competent, and their application depends on the leader's position in the organization. The nine leadership competencies are:

1. **Communication:** There's so much that can be said about communication. It is the "must have" link between the leaders and soldiers that help cause the mission to be accomplished. Communication is effective only if the information is understood between the persons communicating.

    Good leaders always keep their emotions under control and make every attempt to listen objectively. They know they can't respond intelligently to what others are saying until they know exactly how they feel. The best action you can give, when talking or listening to someone, is real and revealing facial expressions.

2. **Supervision:** Supervision lets the leader know if his orders are understood and shows his interest in the soldier and the mission.

   By considering the soldier's competence, motivations and commitment, the leader can judge the amount of supervision needed. Remember, though, over-supervision can cause resentment—and under-supervision can cause frustration. If the leader is not competent in the job or duty being supervised, another competent leader or soldier should do the supervising.

3. **Teaching and Counseling:** One of the most effective ways of improving performance is by overcoming problems, increasing knowledge or gaining new perspectives and skills through teaching and counseling.

   Teaching your soldiers is a sure way of preparing them to succeed and survive in combat. You must take a direct hand in your soldier's professional and personal development.

   Counseling can be a "pat on the back" for doing a good job, or locking his heels to receive a "chewing out". Counseling can be conducted in the motor pool, under a tree, on the FTX grounds or in an office, under more formal conditions. Counseling should influence the solder's attitude or behavior. The leader should be available whenever the soldier requires assistance, but should try to lead the soldier into solving his own problems.

   Personal counseling should adopt a problem-solving approach, rather than an advising approach. The leader may have to refer some situations to another leader, the chaplain or one of the many service agencies available to military personnel, if needed. No leader will be able to solve all problems that a soldier might have.

4. **Soldier Team Development:** Combat is a team activity. Cohesive soldier teams are a battlefield requirement. The leader must take care of his soldier and conserve and build their endurance, spirit, skills and confidence to face the inevitable hardships and sacrifices of combat.

   Before a unit can function as a team, there must be a strong bond between the leader and his soldiers. The effectiveness of a cohesive, disciplined unit is built on the bonds of mutual trust, respect and confidence. Good leaders know how important it is that peers, seniors and soldiers work together to produce successes. Soldier team development is a must in training and orienting soldiers to new tasks and units.

5. **Technical and Tactical Proficiency:** This area is essential to leadership. The leader must be proficient with each weapon, vehicle and piece of equipment in his unit. The next major battlefield will have no safe areas to the rear, the flanks or in the air. The leader may have to organize and lead a platoon of clerks in defense against a well-armed enemy unit.

   The Battle of the Bulge in World War II might have been lost without the heroic efforts of small groups of well led, determined soldiers. Your soldiers need to know that they are following a competent leader who knows his job, and that you, as a leader, are able to train them, maintain and employ your equipment and provide combat power to help win the battle. You must know these things and instill confidence in your soldiers in your abilities.

   Tactical competence requires you to know warfighting doctrine so that you can understand your leaders intent and help win battles by understanding the mission, enemy, terrain, troops and time available. Technical and tactical proficiency work hand-in-hand.

6. **Decision Making:** You, as a leader, should be able to make timely decisions and state them in a clear, forceful manner.

   The wise leader gathers all facts, weighs them, and then calmly and quickly arrives at a sound decision. It's important that decisions be at the lowest organizational level where information is sufficient. The leader's goal should be to make high-quality decisions the soldiers accept and execute quickly. The leader should keep in mind that many sound ideas originate at the soldier level.

   When making decisions and solving problems, the leader must:

   - Recognize and define the problem
   - Gather facts and make assumptions
   - Develop possible solutions
   - Analyze and compare the possible solutions
   - Select the best solution and take action

When time permits, the leader should involve his soldiers in the decision making process. Sometimes they have information or experience that will lead to the best decision or plan. The leader should try to identify the best course of action that is logical and likely to succeed.

7. **Planning:** Planning is usually based on guidance or a mission you receive from your soldiers or higher headquarters. If possible, the leader should involve the soldiers in planning. Their ideas may help develop a better plan and the process also gives them a personal interest in seeing the plan succeed.

   Planning is intended to support a course of action, so that an organization can meet an objective. Soldiers are happy with leaders that keep them informed and plan training and operations to ensure success. Planning involves forecasting, setting goals and objectives, developing strategies, establishing priorities, delegating, sequencing and timing, organizing, budgeting and standardizing procedures.

8. **Use of Available Systems:** In today's Army, all leaders need to know how to use computers, analytical techniques and other modern technological means that are available to manage information, and help the leader and soldiers better perform the mission.

   The leader must use every available system or technique that will benefit the planning, execution and assessment of training. The leader must understand that computer technological advance is important. It could give him and his soldiers the edge on the battlefield.

9. **Professional Ethics:** This area includes loyalty to the nation, the Army, the unit and the duty, selfless service and integrity. The leader should always set the example for his or her soldiers and must learn to be sensitive to the ethical elements of each situation he will face, including orders, plans and policies. The oath every soldier takes, requires loyalty to the nation and involves an obligation to support and defend the Constitution. Loyalty to the Army means supporting the military and civilian chain of command. I will further address this subject later in the book.

WHAT THE LEADER MUST BE, KNOW AND DO

As a leader you must be a person of strong and honorable character. Your visible behavior is an implication of your character. Your character describes, as a sum total, your individual inner strength and is a link between your values and behavior. The leader with a strong character sees clearly what he wants and takes action to get it.

Soldiers are attracted to leaders with strong character. On the other hand, leaders with weak character are not sure of what they want. They lack the willpower, purpose, drive and self-discipline that leaders with strong character have. Their traits are out of order. They hesitate, are not consistent and do not attract soldiers.

A good leader can have a strong character and immoral values, but a great leader will have strong character and moral values. Your soldiers will judge your character as they notice your day-to-day actions. They will learn if you are wishy-washy, lazy, selfish, open or honest. Soldiers will trust their lives to leaders, based on their judgment of whether or not the leader has strong character traits like courage, candor, competence and commitment.

If you have a weak character, your soldiers may still follow you, but only because of your rank and their sense of duty—not because you are a good leader. The more you display traits like integrity, maturity, will, self-discipline, flexibility, confidence, endurance, decisiveness, coolness under stress, initiative, justice, self-improvement, assertiveness, empathy or compassion, a sense of humor, creativity, bearing, humility and tact, the more you will increase your chances for effectiveness in difficult crisis situations.

To build character, you must first be honest about your character weaknesses and demonstrate the self-discipline on which strong and honorable character is based. To build strong and honorable character, the leader should:

- Be honest in judging the present strength of his values and character
- Determine what values and beliefs he wants to promote
- Identify his character in relation to the mission and/or situation
- Imitate other leaders who demonstrate the values, beliefs and character you are trying to develop

Developing character in an on-going process. You must work hard, study, understand yourself and exercise your will power and self-discipline to strengthen your character. As a leader, you must understand and demonstrate loyalty to the nation, the Army and the unit.

Duty, selfless service, loyalty and integrity are the four elements of the Army professional ethic, which are the principles that guide professional soldiers to do the moral and right things that ought to be done. The leader who is loyal to the nation is one who has a deep belief in serving and defending the ideal of freedom, truth, justice and equality.

Loyalty to the Army means supporting the military and civilian Chain of Command. Loyalty to the unit means that the leader places the needs and the goals of the unit ahead of his own.

It's the duty of all leaders to accept full responsibility for his soldiers' performances. They know and earn their rank and leadership positions to serve their soldiers, units and the nation. The leader's three general ethical responsibilities are that he must:

- Be a good role model
- Develop his soldiers ethically
- Lead in such a way as to avoid putting soldiers into ethical dilemmas

Leaders should avoid creating ethical dilemmas for soldiers that may cause them to behave unethically and create trouble for themselves and their soldiers. Some examples of this are:

- Zero Defects
- No excuse for failure
- Can-Do
- I don't care how you get it done—just do it
- Covering up errors to look good
- Telling superiors what they want to hear—not being honest with them
- Setting goals that are impossible to reach
- Showing more loyalty to superiors than subordinates
- Falsifying reports to say what your leader wants to see
- Disrespect flows up—not down

True ethical dilemmas exist when two or more deeply held values collide. In such situations using a decision-making process can help you identify the course of action that will result in the greatest moral good. When making an ethical decision, the leader should:

- Interpret the situation and find the ethical dilemma
- Analyze the factors and forces that relate to the dilemma
- Choose the course of action that he believes will best serve the nation
- Implement his chosen course of action

A variety of forces will influence the ethical decision making process. Before you take action, ask yourself if you could justify the morality of your action before a group of your peers and leaders. Be true to yourself and the principles on which this nation stands and do what you believe or know is right. Some of the factors and forces that may influence your decision-making process are:

- Laws, orders and regulations
- Basic national values
- Traditional Army values
- Institutional pressures

When faced with a situation where the right, ethical choice is unclear, consider all forces and factors that relate to the situation and then select a course of action you feel will best serve the ideals of the nation. Beware of the situation ethic, which is a situation that makes you believe you should do what you know is not right. Taking valuables from a dead soldier or holding enemy soldiers as hostages with the hope of freeing yourself are two examples of this.

The leader should also know about standards. Not only will the leader have to be able to meet or exceed these standards, he will also have to enforce them on his soldiers. CTTs, ARTEPs, regulations, laws, LOIs, orders and training schedules all contain some type of standard that must be met.

Standards do not have to be easy to attain, but they must be attainable. You, as the leader, must be able to attain the standards you set for your solders. If not, how can you be an effective trainer and enforce that which you cannot do yourself? If the standards you set for your soldiers are too high to reach, you will create low morale. It is best to set high standards in stages and have your soldiers reach one stage at a time, than to have them reach for the stars all at once.

You must communicate standards clearly and ensure they are understood, reachable and that the soldiers know what is expected of them. If you are a leader and cannot meet the standards that you enforce, you may be able to live with it, but sooner or later it will affect your career.

Years ago, one of my unit commanders was overweight, by some estimates, over forty pounds. Everyone knew it and it was the number one IG complaint. AR 600-9, at the time, stated that if a soldier was overweight, that soldier could not hold a leadership position. His boss, the Battalion Commander, knew about the weight problem, but did not remove him from the leadership position.

I must admit, my old commander was a good leader. He made sure that the officers were trained and that the NCOs did their jobs. He also spent many hours in the gym trying to lose weight, and, if there were soldiers in the unit that were overweight, he would take them along. He didn't like to run, but, when we did, he was there.

Although he did try to lose the weight and performed admirably in his position, when it came time to PCS, the Battalion Commander, despite the fact that he allowed him to continue in the position of commander, gave him a bad OER, just in case there was an investigation that might come back on him.

In this case, both my Commander and my Battalion Commander were wrong. If the Captain hadn't taken the position, or, when the Lieutenant Colonel realized the problem, removed him from the position, he could have taken another job, worked on his weight problem and still got a good OER. It is important that if you support your soldiers when knowing that they are doing wrong, don't turn around later and stab them in the back.

One of the duties of a First Sergeant is to train the platoon sergeants to become First Sergeant. If he doesn't know his job, however, it will be very hard to train the platoon sergeants successfully. All leaders should be well trained, but junior leaders should learn as much as they can and build on it as they climb to the top. Once you attain a top-level position you must perform at that level or risk being relieved from duty.

The junior leader is the most important of all the leaders and it's important that his soldiers trust, respect and are willing to follow him into combat. Let me explain.

Computer simulators are now being used to determine the senior leaders' abilities to fight and win the battle. Field grade commanders and other senior officers being tested, go into the so called "War Room" which is no more than a room filled with all kinds of computers to run the simulator. The leaders are given situations and it is left up to them and their staff as to what should be done in different situations. Some of the situations they will encounter are: shortages of food or water, supply vehicle break downs or captures, the senior commander is killed, and no communication with higher headquarters. The simulator will win every time unless the commander and staff are good and know how to use the program.

One thing the simulator can't simulate is the feeling, morale and skill of soldiers. If the "computer trained" commanders and staff have to take their soldiers into combat, they are going to remember their simulator training and are going to try to do what they did in the war room. It is that junior leader and his soldiers that will have to do what the computer trained commanders tell them. At the same time, they will be on the battle field with their soldiers, and will see, first hand, what is going on with them.

The junior leader will be the one that knows his soldier got shot before the commander does. He will be the one that sees some of them cry and call for their mothers. He is the one that they will depend on to get them back home to their loved ones. Simulators are no more than a big video game, but the war zone is for real. You may not get to play, again.

THREE LEADERSHIP STYLES

The way a leader interacts directly with his soldiers is his style of leadership. As you examine three leadership styles, you will see that your values and beliefs have a great deal to do with how you lead your soldiers and which style is best for you. Whatever your style of leadership is, it will change due to the many situations you will face. Most likely, however, you will always come back to your style.

Your leadership style will depend a lot on training, education, experience, environment, values, beliefs, your soldiers and your view of the world. Talk to other leaders that have your style and learn more.

Directing Style of Leadership

When you tell your soldiers what you want, how, when and where you want it done and then supervise closely to make sure that it gets done, you are using the directing style of leadership. This style of leadership is good for leaders to use when leading soldiers who lack experience or competence for the task.

On the other hand, this style of leadership may upset a soldier who has experience and is very capable of completing the task. There will be times when only the leader may know what needs to be done and how to do it. At those times he may want to use the directing style, even with experienced soldiers. Remember, at these times, however, that a leader should use tact with the experienced soldier.

On the battle field, the directing style will be used more than any of the other styles of leadership and, there may not be much supervising or follow-up because of the dangerous or terminal properties of the mission. The leader knows the soldier's life will be at stake, but time may not be sufficient for him to pass on all the information he has to the soldier. The directing style of leadership is mostly found at basic training units.

Delegating Style of Leadership

The leader uses the delegating style of leadership when he delegates problem solving and decision making authority to a soldier or a group of soldiers. If a leader uses this type of leadership, however, he should be prepared to shoulder the responsibility, should something go wrong. A leader is always responsible and accountable. Authority to accomplish the mission can be delegated, but, responsibility cannot.

Some things are okay to delegate while others are not. Make sure that your soldiers are capable of accomplishing the task, before you delegate it to them. This takes training and knowing your soldiers. Delegating authority is a chance taking process. It can cause you to be relieved from duty and/or the ramifications or consequences that will follow.

When you receive new, inexperienced soldiers, you may start training them by using the directing style of leadership. As time goes on, however, and they become more competent, motivated and work well in the combat team, you can start supervising them less, encouraging them to ask for advice and allow them to participate in helping you make plans and decisions. With time, experience and your skillful leadership, you will soon be using the delegating style with your once inexperienced soldiers.

The delegating style of leadership is the most efficient of the three. It requires the least amount of time and energy to interact, direct, and communicate with your soldiers but you must ensure that your soldiers are well trained and responsible enough to accomplish the task. Competence is a must when delegating authority to one or more of your soldiers. This style of leadership is found throughout the Army, but it's more noticeable in Combat Service Support type units.

Participating Style of Leadership

This style of leadership is one of the best ways to build your soldiers confidence. With this style, the leader involves his soldiers in the decision making and planning process. The leader will use this style when there is sufficient time to work things out or make a decision with the soldier's input.

The most important thing the leader must remember, when using this style of leadership, is that he maintains overall responsibility for the quality of the plans and decisions, and that he must make the final decision to approve or disapprove those plans prepared by his soldiers.

Having your soldiers participate is a sign of strength that they will respect, also you will be teaching them to make decisions which, in turn, will help them become better leaders in the future. Soldiers care more about making goals and plans that they were involved with work, than those made by their leader or others. If no harm will be done and time permits, it's sometimes better to say, "We'll try it your way, and if it does not work, we'll try something else." Notice I didn't say, "..if it does not work, we'll do it my way." If you say it that way, the soldier will think you want them to do wrong, but, if you say it the other way, then they feel they will get another try, and it's not a do or die situation. If their plan does not work, they will still respect you for letting them try.

The leader must be able to choose the right style of leadership, at the right time, for the right situation. Remember your values and beliefs will influence the style you select, as well as the degree of confidence you have in your soldiers.

UNDERSTANDING LEADERS RELATIONSHIP

There must be an understanding between the leader (you) and the senior leader. The senior leader is the person that you work for. He is your direct supervisor, the one that does your NCO-ER or OER. You have only one boss, and your soldiers only have one boss.

If you are the platoon sergeant, the soldiers do not work for you, they work for the section chief, or, in some cases, the squad leader. If you have soldiers that work for you and you don't take charge of them, someone else will. When that happens, they will have two or more leaders they work for. You should be the one that directs your soldiers. You should be the one that says if they can go on leave or pass. You should be the one that says whether or not they get promoted.

There are many cases where the senior leader will have a say, also, but you should be the first one to say yes or no. No other leader should tell your soldiers what to do, without first checking with you, and, if they do have them doing something, they should contact you and let you know.

PFC Clark was assigned to one of the Hawk missile batteries in West Germany, and, like most good soldiers, he always did what he was told to do by the leadership in the unit or in other units. SGT Taylor was his crew chief, and SSG Evans was the section chief. His platoon sergeant was SFC Brown and the platoon leader was 1LT Lewis.

One morning, after work formation and police call, SGT Taylor told PFC Clark that he wanted him to perform PMCS on the section's loaders (used to transport missiles) and have them dispatched. When PFC Clark got to the maintenance section, he saw his section chief (SSG Evans), who told him to go back to the unit to sign his leave request at the 1SG's office, and, if he hurried, he would have time to catch the unit bus that was taking the old manning crew back to the unit.

After the ride back to the unit, PFC Clark had to wait for the 1SG, who was with the commander, in her office. When the 1SG returned, PFC Clark signed the leave request and went back outside to get a ride back to the tac-site. His platoon sergeant (SFC Brown) saw PFC Clark waiting for a ride and gave him a lift back to the site.

On their way back to the section, 1LT Lewis saw the two of them and told them he wanted them to go back to the unit with him, pick up supplies for the section and that he would not be able to return with them. By the time the supplies were picked up from the supply room, it was time for lunch, so SFC Brown and PFC Clark went to lunch.

During lunch, the battalion Organization Readiness Evaluation (ORE) team arrived on site to check the crew and missile system. SGT Taylor's section did not pass, because neither loader was dispatched and one was low of oil. The commander was very upset about this, and told 1LT Lewis that she wanted to see him and his crew on site after duty hours.

After the meeting, about 1800 hours, 1LT Lewis told the crew that they could not go home and had to stay on site until they passed a battalion ORE, and that PFC Clark would receive a three day pass, because the commander said that he must need a rest after working for so many leaders. This is an example of why there must be communication between leaders.

There have been too many times that leaders have been relieved from duty, because one didn't know what the others were doing. All leaders must understand the senior leader's strengths, weaknesses, work habits, leadership style, values, beliefs and standards. If theirs are the same, or close to the same, as the senior leader, then they can build an effective team. If not, the leader must adapt to the senior leader's ways.

There must be an agreement as to who will do what and how they are to support each other. Each must understand that the combined competence of the platoon or section they supervise is the deciding factor of their success as the leader and senior leader. Once the leader understands the senior leader, it's possible to form a good working relationship, even if the leader has to adjust his ways to that of the senior leader.

The best leaders are those who can communicate and work effectively with their seniors. The leader should do his best to give the senior leader what he wants, but, at the same time, he must use tact. He must say what is on his mind and be able to explain the reasoning behind the actions he takes and suggestions made. At the same time, he must remember that the senior leader's final decision should benefit the platoon or section, as a whole.

The leaders must understand that the two of them, as a team, are in charge of the soldiers, yet the senior leader is responsible, overall, for everything. If there is no respect or togetherness between the leaders, members of the team will lean towards the one that will benefit them more, or seek personal help from other leaders. The responsibilities of the leaders are many, but they can be accomplished with ease, if there is a good relationship between the leader and senior leaders.

2

KNOWING AND TRAINING YOUR SOLDIERS

PUTTING YOUR SOLDIERS FIRST

As a leader, your soldiers should be your number one concern, not yourself, your equipment or your mission. Now, before you throw this book into the trash, let me ask you three questions. Who operates, cleans and takes care of your equipment? Who accomplishes your mission? Who is the main reason you get promoted and get a good NCO-ER or OER? The answer to all three questions should be your soldiers.

Without the soldier, the Army doesn't need you, and without them there would be no Army. Your primary duty, in the Army, is the care and welfare of your soldiers. You, the leader, are their mother and father of the military world.

I don't care what type of "High-Tech" equipment you have, or how many smart bombs or computers you may use for fighting the enemy, you still need your soldier. If you don't have any soldiers, then you are someone else's soldier. Leaders get promoted and get good evaluation reports because of their abilities to train and lead soldiers. I know that leaders are soldiers, also, but when I refer to soldiers, I am referring to the junior NCO and below.

There are still many senior leaders who will tell you that the mission comes first, and then the soldier. That's okay—that's the way they were taught, and they teach the same thing. Remember, it's not easy to change someone's beliefs. There are still leaders who say that females shouldn't be in the Army, yet 35,000 of them went to the Gulf. Those same leaders will be the first to try on new uniforms, but they can't deal with new changes. Take care of the soldiers, and they will take care of the mission when the time comes, whatever the mission may be.

Did you know that most Army equipment gets more attention than the soldiers? Think about it—they will spend almost a full day or days in the motor park, pulling PMCS on the equipment. Then, look at how long it takes to inspect them in the morning formation, if they are inspected. Senior leaders will go into a motor park, and the first thing they will do is check the fuel levels, yet they don't know if the soldiers living off post have food to eat or not.

If a soldier is having a problem at home or needs food, most of the time they won't say anything, because they think their leaders don't care—and in many units, they are correct. If you want to be one of the best leaders, get promotions, good evaluation reports and have the respect of your soldiers, make the soldier number one on your list.

INFLUENCING YOUR SOLDIERS

The only way that you can influence your soldiers to do something is by making them want to do it. If you want good performances, you have to know how to talk to them and be skilled enough in using tact that they will not only want to do it, but will feel good about doing it. There are many ways this can be done. Four of the best ways are:

- Make them feel important
- Show appreciation and encouragement
- Become interested in your soldiers
- Give them what they want

We all want to feel important. We want others to care about us and we want to do things that make them think we care. Many teens are in gangs today, because they feel more important there than they do at home. Unlike at home, they feel needed in the gang, even though they may do things that are wrong.

Robin Hood was wrong, but many people liked him, so he kept on taking from the rich and giving to the poor. There are many people that will rob, steal or even kill, just to see their name in the newspaper or on the six o'clock news. There are many things that make some soldiers feel important. Some are:

- Getting awards
- Getting a pat on the back
- Becoming soldier of the month, quarter and/or year
- Getting promoted
- Time off
- Conducting PT or in charge of the run
- Being put in charge
- Having the best military bearing
- Having the best display during inspection
- Marching the section, platoon or unit
- Being mentioned in the unit or battalion newsletter

Some soldiers will go all out to get things that make them feel important, and, after they get it, they will try to get the next thing that makes them feel important or do the same thing, again. They will also talk about things that make them feel important. Listen to them and what they are saying.

To many soldiers, a pat on the back or a "Thank you" means more to them than a promotion. Too many leaders will jump all over the soldier when he does something wrong, but give them no credit when they do something well. Lack of appreciation is not only the reason many loved ones leave home, it is also why many soldiers go AWOL. Award your soldiers, let them know you appreciate them. It's not as if you are giving them something for nothing. What one junior NCO at Fort Hood did, when his troops returned from the Gulf, is one example.

This particular SSG and his soldiers were scouts for an armored battalion and were far ahead of the main line. When they were called back to the rear, one of the tanks threw a track, so they had to stop and fix it, or leave the tank behind. Knowing that the enemy was in the area, they set up a defense and, at the same time, began working on the track. They were able to get it repaired and move out of the area, just before the enemy arrived.

Because of the situation and his soldiers' actions, the SSG put his soldiers in for an award, but it was denied. Later, the platoon leader, again, tried to get them awards, but, again, the awards did not go through.

The SSG told the soldiers that the awards were denied, but he knew that he had to do something for them. He went down to a local print shop, had some 11"x14" certificates of appreciation made up, and took them to an artist, who drew pictures on each one showing what each soldier was doing at the time the track was being repaired. When they were done, he and the platoon leader signed them, and had the commander present them to the soldiers in the morning formation.

I will be willing to bet you that those awards meant more to those soldiers than any other award they received. That SSG, in my book, should have been the CSM. Not only did he know how to influence his soldiers, he knew how to show his appreciation. Honest appreciation gets results where criticism and ridicule fail.

Criticism will not help you influence your soldiers. It is ineffective because it puts them on the defensive and usually makes them strive to justify themselves. It is also dangerous, because it wounds their pride, hurts their sense of importance and arouses resentment. Criticism must be done with tact to be effective.

Too many senior leaders criticize junior leaders for some of the same mistakes they made when they were junior leaders. That's okay, but the junior leader would feel much better if the senior leader said something like "I remember doing that same thing when I was a _____. To correct it I _____ and, before I knew it, all was Okay again." Not only will that make the junior leader feel better, but, knowing that you did the same thing will make them feel like they are on the right track.

You must also take into account that the junior leader is trying to get where you are. The way you train your leaders is, most likely, the way they will train theirs, and remember that a sorry leader will create more sorry leaders. Train junior leaders correctly, and they will do the same towards their soldiers.

SGT P. was one of the crew chiefs in the platoon, and he was in charge of keeping the G.P. Medium tent warm for the soldiers. Whenever the platoon went on an FTX, I always had a list of things that had to be done before, during and after the FTX. I would assign an NCO to each task that had to be accomplished.

When I placed SGT P in charge of the heater in the tent, all he had to do was have one of his soldiers place two five-gallon cans of fuel outside the tent, for the heaters. During the night, if the fuel ran out, his assigned soldier or one of the guards would replace the empty can with a full one.

About 0200 one cold winter morning, I was laying on my cot, looking at the flames from the heater die down. I never could sleep well in the field, maybe because I always went to bed when the sun went down, but,

I think the real reason was because I always wanted to get my soldiers back home, safe. It was nothing to see me, early in the morning, talking to one of the guards or looking around the area, seeing if all was okay. I also knew that it was my job to train my leaders to become platoon sergeant one day, and being an NCO is a day and night job.

The flame from the heater got smaller and eventually went out. When it did, I called SGT P, who was sleeping about two cots over from me. When he looked up, I told him that the fire was out in the heater. He called for one of his soldiers, but I told him, "No, SGT P, I told you to change the can."

Without saying another word, he sat up and started getting dressed to go out, in the cold, to get the fuel. I could hear him, outside, changing the cans, and watched him come back inside and light the fuel in the bottom of the heater. He pulled his cot closer to the heater, and just sat there, getting warm.

Later that day, he asked me why I had made him change the fuel cans. I told him that when someone is in charge of something, they are responsible for it. They can pass on the particular duties, but they are still responsible for making sure that it gets done. I said, "You didn't make sure your soldier did the job, so I had to make sure my soldier did the job. If not, the platoon leader could make me do it."

Another way to influence your soldiers, is to give them what they want. If you do, you will get what you want. Every act you have performed was because you wanted something. Your soldiers are the same way.

One thing I did, when I was a section chief, was get with my section after the morning formation and let them know what had to be done that day. I always had more to do then they normally would do in one day. I would read down my list and ask who wanted to do it. Whoever raised their hand and got the task knew that, when they completed the task, they were off for that day. Most of my SPCs and below were gone by 1400 each day. The junior leaders and myself would stay around until close of business, but sometimes I would give them time off, too. Because I was able to do that, my section got more accomplished and there was always someone around.

Another thing I did as platoon sergeant was, after an FTX, when we would return to the unit, the platoon was divided into many sections to clean and repair the field equipment. It didn't matter what time we returned to the unit, we would start cleaning and repairing because the soldiers knew that they were off when they were done—sometimes as much as a day-and-a-half. You see, we had three days to get things back in order, so whatever part of the three days weren't used was time off.

The platoon leader, myself and two or three of the junior NCOs were always there, so, if something was not done or something additional had to be accomplished, we did it (and when I say "we", I was included). Soldiers have a great respect for leaders who don't mind getting their hands dirty, because they believe that a leader who will work with them, will fight with them, should they go to war. As you read this book, you will find many ways to influence soldiers.

GETTING TO KNOW YOUR SOLDIERS

To know yourself is to know what you are capable of doing. To know your leader is to know what he expects of you and the support that you will receive. To know your soldiers is to know what they can and cannot do.

Knowing your soldiers is much more than the information you may have about them on a 3x5 card. It's knowing what makes them tick—their goals, values, beliefs and wants. The more you know about your soldiers, the better your chances will be to influence them in accomplishing the unit's mission. It's important that you are aware of your soldiers' attitude, because attitudes are no more than strong beliefs or feelings toward people or situations.

Attitudes are not quick judgments, they are acquired throughout our lives and are part of our personalities. We all know that most of the time a positive attitude will result in a positive conclusion. Your soldiers' attitude, about many things, will not be the same as yours. You may like inspecting your soldiers, but they may not like being inspected. You may like running them five miles, and they may like running only two miles.

To change their attitude about the inspection and get a good outcome, give them something they all like—maybe the day off, after the inspection. You may also tell them that there will be two inspections, but only one, if the outcome of the first one is good. You may run two miles for three days, or five miles only one day. Once you learn to meet them half-way, you will find that you will be able to get more done and a better outcome each time.

Sometimes, a leader will think a soldier has a bad attitude, just because he feels different about a given situation. Your soldiers will shape their attitude about you by what they see and hear. They will interpret your attitudes through your behavior. Attitude represents a powerful force in any organization. Leaders will generally encourage certain attitudes and punish others, so soldiers will tend to develop

KNOWING AND TRAINING YOUR SOLDIERS 31

attitudes that minimize punishment and maximize rewards.

Soldiers pay special attention to the behavior of their leaders. Therefore, if a leader leaves the job, before it's time to get off, the soldiers will soon start doing the same thing. They will develop the attitude that staying until quitting time is not important. Leaders of the unit have the greatest impact on the soldiers' attitudes. Soldiers pay more attention to what their leaders do than to what they say, so leaders must demonstrate the kind of behavior they want the soldiers to develop. All leaders are some soldier's role model and must be aware of that fact.

As a leader, you must realize that you only have control of your own attitude, and can change it at will. To change your soldiers' attitude, you must produce an atmosphere in which they will want to change their thinking. The two most powerful ways you can lead others to adopt the attitude you wish them to have is; (1) affecting the consequences they will receive after altering their behavior toward the attitude you want them to embrace and (2) changing the conditions that surround the situation soldiers want.

An example of affecting the consequences is, let's say, if you want your soldiers to max the PT test, reward them for their efforts. As a rule, soldiers will willingly do things that bring about positive consequences for themselves.

Soldiers, also, want to work with leaders that treat them with respect. They want interesting work, recognition for good work, the chance to develop skills and the opportunity to be promoted. Satisfying working conditions lead to positive attitudes.

To know your soldiers is to know their wants and needs, and be willing to help them accomplish what they want to accomplish. If you spend as much time with your soldiers as they spend in the motor park, you will know them as well as they know the equipment. The best time to start learning about your soldiers is as soon as you or they report to the unit.

When you report to a new unit, the soldiers will judge you by how you dress and carry yourself. Their main concern will be if you are able to lead them in a way that they can accomplish their personal and unit's goals. If you are replacing a good leader, they will expect you to be the same way, or better. If the leader you replace was not a good leader, they will expect you to be better. Don't you feel the same way about your leaders?

Soldiers don't care about a leader being hard core, as long as he is fair and is consistent. Don't try to be the leader you are not, because you will not be able to change overnight. If you must make adjustments, make them a little at a time. To be a well received leader, you must be able to meet or exceed the standards set for your soldiers, and your soldiers will always expect you to out-soldier them.

After reporting to a new unit, meet with your soldiers, or should I say, the soldiers you will be in charge of. If you are a First Sergeant, you will want to meet with the platoon sergeants, first, and then the unit soldiers. If you are a platoon sergeant, the section chiefs should be your first concern. The section chiefs will meet with the section, but the leaders should always be first on your list.

When meeting with your soldiers, start your meeting by letting them know what you will like to do for them. Talk about time off, leaves, passes, awards, promotions, caring for the family and other things that are important to the soldiers. After that, let them know what you expect of them and ask if there are any problems. If the soldiers have another first line supervisor, let them know that they should let their leader know the problem, and let their leader know that you are there if he needs help. Never try to take over another leader's responsibility.

Too many leaders go into a new unit and try to get the soldiers to like them, for fear that their leadership shortcomings will be shown. Give yourself time to find out what is going on, before making the big decision. Try not to pre-judge your soldiers, as they will do you. Know them for what you know about them, not for what someone tells you about them. The sooner you get to know your soldiers, the sooner you can start leading them in accomplishing the unit mission.

When a soldier reports to your unit or section, it's important that you, the leader, provide him with all the help or assistance needed. The sponsor should be someone who will be with them each day, until they are completely inprocessed into the unit. Use your best soldiers for sponsorship, not someone that's about to PCS, even if he is a good soldier.

SOLDIER'S VALUES

The development of the four basic values (candor, competence, courage and commitment) in each soldier can help strengthen the acceptance of the values of the Army ethic.

Candor is honesty and faithfulness to the truth. Team members must be able to trust one another and their leaders. Without truthfulness, this will not occur. When soldiers see their leaders or peers lying about status reports or other situations, they wonder if they can be trusted to be truthful in a crisis. The question arises, "Will they be honest about the wartime situations?" There is no time for second guessing in combat.

Competence is imperative for the combat-ready team. Soldiers accept one another and their leaders when they are satisfied with their leader's and peer's knowledge of the job and their ability to apply that knowledge in the working situation. Nothing deteriorates teamwork quicker than the perception that soldiers do not know how to soldier, and leaders do not know how to lead. The soldier's competence is the basis for the self-confidence critical to feeling accepted by the team.

Courage, both moral and physical, is displayed by soldiers in cohesive, combat-ready teams. They understand that fear in combat is natural and can be expected. Moral courage helps the combat-ready team to do the right thing in a difficult situation, even when some might strongly feel that the wrong is more attractive. Courage on the part of one or two soldiers is contagious and becomes a way of life in the cohesive, combat-ready unit.

Commitment to the unit, the Army, and the nation occurs when soldiers accept and demonstrate the values discussed above. When soldiers willingly spend extra time to get the job done for the unit, they show that unit accomplishment takes priority over personal inconveniences. They are demonstrating commitment to the unit and to the Army.

These values apply in sections, squads, crews, platoons and companies, and, it's up to the leader to ensure that they are kept intact. Years ago, when I was stationed at Fort Bragg, I encountered a situation that I still find hard to believe, today.

It all started, one day, during the morning formation when the platoon sergeant was talking to the soldiers about something that had gone wrong the day before. He said something like, "If you mother-_____s don't get your heads out of your _____, I'm going to put you all in jail." I just stood there, finding it very hard to believe.

After the formation, I went up to him and told him that that was no way to speak to soldiers. He replied that he knew, but they had just got to him that morning. Well, I didn't fall for that. What he said was one thing, but I didn't think he was a good leader, anyway, because of his military bearing. He, also, didn't do things that a platoon sergeant should do.

About a week later, I told him that I was going to ask the commander for his job as platoon sergeant. He and I were both SFCs, and either had the MOS for platoon sergeant, but my DOR was higher than his, and we were in the same platoon. I spoke to the commander and the first sergeant, and got it. He was sent to higher headquarters and placed on the evaluation team, the place I think he wanted to be, all along.

I had my work cut out for me, because the soldiers were having their way about things. Anyway, to make a long story short, it all worked out well for the unit as a whole and the 1SG wanted me to receive the Meritorious Service Medal for my performance. I told him that I didn't think I deserved it, because I didn't think that a person should get that medal for working a year as a platoon sergeant, so I was given an Army Commendation Medal. I'm not trying to say that I was the best leader in the Army, but I was one that cared about the soldiers.

You may say that I was lucky to have had leaders who cared about me, when I was coming up in the ranks, and I just passed it on. You don't have to be the best leader to know something is wrong and try to correct it, you just have to be a leader who cares and wants to do what is right.

INGREDIENTS FOR A COMBAT-READY TEAM

A team, outnumbered and overpowered, can overcome lack of strength and win, when it has a strong desire to do so. That strong desire is called spirit, a most critical ingredient of a combat-ready team. Soldiers in a unit with spirit, believe in the cause for which they are fighting. They believe in themselves, and they fight for one another.

Professional soldiers are mature and share the values of their profession and their unit. A mature soldier develops physically, socially, emotionally and spiritually. Physical maturity provides the stamina necessary for sustained action and intense stress. The soldier that falls behind in the unit PT run may very well be left behind in combat. There is no way that a soldier in a PT uniform, who falls out of the two mile run will be able to run two miles in full combat gear. That's why many units run more than two miles.

Social maturity provides the willingness to work with others in cohesive teams. Soldiers are willing to work with their peers and leaders. All they want is to be able to do their jobs and be treated with respect and given recognition for their deeds.

Emotional maturity gives stability to deal with the stress of combat. This is something that only a professional, caring leader will be able to accomplish for the soldier. He must be able to control the stress and fears of the soldiers during combat. The soldiers, however, must trust and believe in the leader, before this can be accomplished.

Spiritual maturity gives the soldiers hope and purpose to face the dangers and uncertainty of combat. Signs of maturity that are important in combat-ready teams include self-discipline, initiative, judgment and confidence.

Self-discipline enables clear thinking and reasonable action in the moment of combat, with its isolation, high leadership casualties, continuous stress and need for independent actions. Self-disciplined soldiers realize that success and survival depend on working together, and the ability to undergo extreme hardship to achieve team goals.

Initiative and judgment are essential in both peacetime and combat. During combat, soldiers need initiative to move decisively in accomplishing their mission. Initiative does not mean "do something, even if it's wrong." It must be tempered by good judgment. Soldiers with initiative, tempered by good judgment, act on their assessments quickly and decisively, with little or no supervision.

To remove doubt and anxiety in combat, the soldier must first have confidence in his or her professional ability, that of the team members and other supporting soldiers. The soldiers need to feel confident in their leaders. The leaders earn their soldiers' confidence as they demonstrate their ability to do their jobs. Soldiers and leader develop mutual confidence by sharing difficult, challenging and realistic training, as well as the rigors and dangers of combat. Mutual confidence multiplies combat power, as it welds individuals into cohesive teams.

The professional Army ethic is also a part of the BE characteristics of combat ready teams—loyalty to the nation, to the Army and the unit. If the leader shows loyalty to his soldiers, he earns their loyalty. They follow legitimate orders, without explanation, because they have confidence in their leader. Soldiers with a sense of duty accomplish tasks given them, seize opportunities for self-improvement and accept responsibility for their actions.

Selfless service is evident in the cohesive, combat-ready team; soldiers and leaders operate with the view that "we're in this thing together." Integrity is the cornerstone of the professional Army ethic. It involves honesty, but, more than honesty, it is a way of life. Trust and loyalty will more likely develop in a unit where integrity is an accepted way of life. The soldier who trusts his leaders integrity, follows his orders, willingly, even in the heat of battle.

The key knowledge that is necessary for the effective teams are: soldier, battlefield, ethical and people knowledge. All soldiers, regardless of military occupational specialty, must master skills necessary for survival in combat. Each soldier is trained to do certain tasks that, when combined with tasks of other soldiers, accomplish the objectives of the commander.

For a unit to be cohesive and combat-ready, soldiers must know what to expect on the battlefield. Soldiers want to know as much as possible about the enemy and the battle environment in order to anticipate the enemy, make decisions quickly, favorably exploit the terrain and win the battle. Soldiers in a cohesive, combat-ready unit take pride in successfully accomplishing their mission with honor. Soldiers and leaders know one another, realizing that others have similar fears and needs, help each soldier overcome his own fears and assist unit members in creating the necessary spirit and "oneness".

Assessment, communication, decision making and training are the key actions performed by soldiers and teams in units of excellence. Teamwork assessment is critical for an effective, combat-ready unit. Most leaders gather impressions by listening, observing and monitoring soldiers' problems. The assessment process is continual. Units grow and change, leaders come and go and the uncertainties of combat impinge on unit teamwork, and, consequently, on combat readiness.

Clear, uncluttered communication is especially critical for teams on the modern battlefield. To operate within the commanders intent, soldiers and leaders in combat-ready teams practice both verbal and non-verbal communication. Decisions are made at every level of the Army. The use of the Chain-of-Command allows these decisions to take place at the proper place and time.

Leaders in cohesive units give team members decision making responsibilities. When this is done, the soldiers learn to do the right thing within the command. The teamwork necessary for cohesive, combat-ready teams requires training. It involves mastering set procedures, such as battle drills or map reading. Teams will more likely react without hesitation in combat, when they have practiced what to do in realistic training. It is important to remember that all units go through a fairly well-defined process of development, which is:

- Building the combat-ready team
- Training the combat-ready team
- Sustaining the combat-ready team

BUILDING THE COMBAT-READY TEAM

The first events in the new soldier's life in the unit, make critical and lasting impressions. Good impressions, created by an effective reception, begin to build the soldier's trust and confidence in his new team.

KNOWING AND TRAINING YOUR SOLDIERS

Initially, the new soldier is concerned about fitting in and belonging. He wants a place on the team, but is not sure how others will accept him. The soldier goes through a process of checking out other soldiers and the leaders. Each soldier adjusts to this new experience differently. Some soldiers "come on strong", bragging about past exploits or telling "war stories". Some adjust quickly, while others fit in more slowly. Some adjust by withdrawing and watching quietly, until they begin to trust others in the unit. Occasionally, a soldier is not able to adjust to the team at all, but all soldiers go through some concern about whether or not they belong in the unit.

Leaders can help new soldiers by realizing that each is searching for some answers to basic questions concerning the teams activities. Leaders also understand the soldiers' concerns, as they attempt to become team members. With these questions in mind, the leader develops a systematic reception and orientation program, designed to ease the new soldiers' transition into the team. It is not enough for the leader to simply give each soldier a quick in-briefing and assign him to a sponsor and a duty position.

I remember one time, when I was assigned to a new unit in West Germany, the unit was on an FTX at the time, but I was taken to the field to see the 1SG. All he did was ask me my MOS, told me who I would be working for and said he only had a SGT for a sponsor (I was a SSG), because the unit was in the field. My sponsor was also getting ready to PCS back to the States, and his main concern was getting his things shipped and his quarters clean.

About two days later, when the unit returned from the field, I went to see the Warrant Officer. It was lunch time, and he and some of the NCOs were playing cards. He looked up at me and asked, "Do you know how to play spades?" When I told him that I didn't, he smiled and said "You must not be a good maintenance man." I knew then, for sure, I had to get out of that unit.

To me, that unit didn't have a reception program. If a SSG was received the way I was, then a private didn't have a chance. The leader must take the time and effort necessary to coordinate the reception of the new member into the team. The leader should spend time talking to the new soldier, in a systemic manner, for weeks after the initial reception, to ensure that he is developing the appropriate goals and understands how his actions contribute to the overall performance of the team.

The leader should check on the new soldier to see if there is a problem with fitting in or abiding by formal and informal rules, such as, meeting appearance standards, yelling the unit's mottos when saluting, etc. Time spent by the leader, on these activities, helps the soldier become a functioning member of the team. Leaders know that most soldiers have an initial desire to contribute to the team, and they build on this personal motivation, realizing that each soldier is different.

As the soldier becomes more competent and reliable, leaders reward their achievements and give them more responsibilities. The leader who looks for positive contributions and gives praise, who takes the position that every soldier is a good soldier until proven otherwise, and who understands that each soldier has strengths that can fit in with the strengths of others to form a strong team, establishes a climate for success.

If the soldier has personal problems, caring means that the leader strives to assist him in dealing with them, whether they are uncovered during the reception or occur during the stay in the unit. Assisting soldiers in dealing with their feelings and concerns, not only reflects caring leadership, it also enhances their effectiveness. Leaders at all levels must take the time necessary to properly receive and orient the members of the team toward the agreed upon goals.

Well developed reception and orientation activities make the leader's team building efforts more efficient. A well-planned reception is an important first step in creating a cohesive team, and it should address the needs of all soldiers. Personnel and administrative problems associated with moving into a new unit are handled promptly and successfully by assigning sponsors to assist new members. The leader tells the sponsor exactly what is to be done and that being chosen as a sponsor means he represents what members of the team should look and act like.

The unit leader should strive to develop bonds between the family and their identification with the unit. It's important that family members understand the service member's duties and the unit's mission. They should have the chance to learn about the unit, it's history and available services and benefits. A well informed family is usually more willing to make the personal sacrifices required to adequately support the soldier and the unit. A family's attitude toward the Army is often based on how the leader treats the soldier and his family. The more welcomed the family members feel and the more informed they are about the unit, the more likely the soldiers will commit themselves to the unit's goals and missions.

Early in the soldier's time in the unit, after accomplishing most of the administrative and family details, the soldier goes through an orientation process. Orientation varies from one unit to the next, depending on the time, leadership and situation. It may be done in a group or individually.

Some important areas to cover in the orientation include:

- Unit values and standards
- Unit mission and goals
- Unit standard operation procedures
- Unit heritage

The leader begins to communicate the values and standards of the unit during the orientation process and reinforces them often during the soldier's stay in the unit. Values such as courage, candor, competence and commitment are communicated to the soldiers directly or indirectly, during the initial training. Such values are not an attempt to change the soldier, but to instill in him the values that history has proven necessary for developing cohesive, combat-ready teams.

If soldiers are going to become productive team members, they must share the values that enhance team performance. The leader transmits these values by example, but may communicate three or four of them during the orientation.

Standards are those principles or rules by which behaviors and tasks are measured as successfully accomplished. Some standards of behavior to which soldiers are held are; saluting, promptness, proper haircut and proper wear of the uniform. These standards support the values of discipline. They are important because they encourage self-discipline in the soldier.

When members of a squad, section or platoon share these values and adhere to the standard that flow from them, they are a more cohesive team. The leader clearly states the standards of the unit to the new soldier, and ensures that the standards are simple, easy to understand, attainable and support the goal of combat readiness. Once soldiers know the standards, the leader is responsible to enforce them fairly, through both rewards and punishment.

Soldiers who develop discipline and live up to high unit standards deserve awards. The goal of both rewards and punishment is to enhance teamwork and thus combat readiness. It's important to remember that unit values and standards are not developed in a vacuum, they need to conform to those of the parent unit and other units with which the soldier works. The leader needs to explain the importance of high standards so that soldiers can take pride in meeting or exceeding them.

The soldier's contribution to mission accomplishment is learning, practicing and becoming proficient in his job. In every case, duty expectations should be related to team accomplishments. The leader's task is to lead each new member to identify with and become a contributing part of the unit.

Combat presents unique challenges to team building. It's basically the same, but combat alters the way it is accomplished. The leader must consider the different dimensions that combat introduces. Some are:

- The time to receive the new soldier is compressed
- The space in which things happen is altered
- The soldier's concerns and feelings are different
- The environment is different
- The way information is transmitted

When the new soldier joins the unit in combat, he comes with a variety of questions. He is thinking of things like:

- What will the people be like?
- Can I trust the leaders and other soldiers?
- Will they accept and care for me?
- What will my job be like?

A positive reception and welcome to the unit will help the soldier feel secure. The unit commander should greet the soldier personally, if possible. He should reassure each soldier that he will be taken care of in all areas, including mail, medical care, and evacuation in case he should be wounded. The 1SG should explain the unit's SOPs to the soldier. The platoon sergeant should explain the platoon's SOPs and basic information the soldier needs to know to work and survive in the platoon. The squad leader should cover the numerous details the soldier needs to know to operate within the squad.

The squad leader is the key individual involved in successfully orienting the new soldier, because he probably has a more direct influence on the soldier than anyone in the unit. The squad leader must also be alert to the many thoughts and feelings that are probably churning inside the new team member. He should encourage the soldier to talk about his concerns, and stress the fact that the squad works together, and that he must do his job well, in order to protect other squad and team members.

The squad leader can use other experienced, positive soldiers in the squad to work the soldier into the unit. It's critical that the members of the squad personally welcome the new soldier and help him "learn 'D' ropes". When the soldier is ready to pull his weight, he and another soldier will become a buddy team.

The soldier will work with his buddy, and, at the same time, actively function as part of the larger squad team. Buddies will intimately know each other and pick up cues from one another, just by watching. A properly selected buddy team causes several positive things to happen. First, the soldier begins to develop close ties of loyalty and friendship to his buddy and other squad members. He sees how he and his buddy are part of the team effort at fire team and squad levels and he will develop a strong sense of commitment to his unit from the bottom up; buddy team, fire team, squad, section and platoon. If the Chain-of-Command has done this right, the new soldier will rapidly become "combat smart".

When the soldier and unit emerge from this building stage, they are ready to further develop into a training stage to develop the cohesive, combat-ready team. The leader must do all he can to help the new soldiers become part of the team. Some of the things he can do are:

- Listen to and care for the soldiers
- Reward positive contributions
- Set a professional example
- Develop reception and orientation procedures for soldiers and families
- Communicate unit values, mission and heritage
- Reassure the soldier with a calm presence
- Provide a stable unit situation
- Talk with each soldier
- Assist soldiers to deal with immediate problems
- Communicate survival/safety tips
- Establish a buddy system

TRAINING THE COMBAT-READY TEAM

The leader must be ready to use the new soldiers as soon as possible after reception and orientation, there he must take the initiative, and get the soldiers involved in the team's day-to-day activities as soon as possible.

There is no clean break between the building and training stages. The processing time depends on leadership, the nature of the group's task, the member's personalities and abilities, and the goals of the team. During this stage, the soldier tries to determine just what he can expect from the unit and the leader, as well as from the team members.

The issues involved in this stage of development will not be restricted to the work place, or to dealings between team members. As the soldier sees his goals and needs being met within the team, he begins to depend on other team members, and they on him, increasing the level of trust.

As members of the team begin to depend on one another, cohesion develops. As the leader detects signs of team growth, he guides the developing team towards mission accomplishment—**like I said, the soldier first, then the mission.** The leader must listen and respond fairly to criticisms or questions, while retaining a firm grasp of the situation. By listening, the leader can discover the soldier's individual needs and can attempt to guide him into accepting team goals.

As he observes and listens, the leader can increase his knowledge about the strengths of individual soldiers—what they like to do and what they do well. He can place them in the job they do best. To deal with possible conflicts over team member's responsibilities and goals, the leader needs to establish clear policies about who has what authority, and under what conditions each team member can exercise authority or make decisions for the team. The leader should explain that, as the new soldiers gain knowledge and experience on the team, their responsibilities and authority will likely increase.

In preparing for combat, all team members must know who is responsible to take over if the leader becomes a casualty. Soldiers look to their leadership to establish goals for the unit. They want a positive direction that will challenge them and provide a chance for reaching their potential. The leader needs to sit down with the soldiers and find out what they expect from the team, both personally and professionally.

If a soldier perceives that his needs are not important to the leader, the process of training a cohesive team will seriously bog down and may never advance to more productive stages of training. A personal discussion between the leader and the member serves five important purposes:

1. It establishes communication between the leader and members of the team
2. It lets the soldiers know what goals can realistically be achieved through membership as an active team member
3. It helps the leader know more about the soldiers and their needs
4. It establishes clear goals throughout the chain of command that are achievable and supports the goals of the higher headquarters

5. It assures the soldiers that their individual thoughts and feelings are at least being considered by the team and it's leadership

Leaders should have plans for getting the soldiers together from time to time to check on their progress. Training is the heart of soldier team development, and all unit tasks and missions are training opportunities. Cohesive teamwork is developed through training activities that motivate and challenge team members. Training is one of the most significant ways the leader can show that he cares, by being concerned enough for the soldier's safety and survival in combat to provide tough and challenging training. The only way to develop teamwork is for team members to do things together. When they work together to accomplish the mission, soldiers experience a deepening sense of unit identity. The most tangible benefit of training, however, is the realization by all soldiers that the unit is either combat ready or close to that goal.

In training, small teams should be given as much responsibility as possible. To achieve maximum cohesion, training goals and objectives must be defined as unit goals and objectives. Training must prepare the unit for combat self-confidence during stressful times, ability to control fears, communication in combat and initiative in the absence of orders need to be an integral part of the training environment. One thing that leaders have at their constant disposal is the opportunity for challenging and realistic training. With that statement, let's go back to West Germany.

Every Thursday we had what was called "prime time training"—at some units it's called "Sergeant's Time". It is the time set aside, each week, when sergeants are supposed to train their soldiers. After about two months, or so, the training became boring to the soldiers. It was boring to me and many of the other leaders, also. As time went by, more leaders started shying away from the training, and soon some of the soldiers were doing the same thing.

I was the platoon sergeant at the time, and I would go around to the different stations to see how training was going. One thing I found at each station was that some soldiers were not paying attention, and there were some leaders that didn't do a good job on the training they were giving to the soldiers. The training was not challenging or realistic, so it became boring to the soldiers. Then it happened.

I was standing at the site gate, one morning, waiting for the guard to let me in. As I did, I looked around and I could see a road going into the woods on the other side of the street, and something told me to check it out. I told the guard that I was going across the street into the woods, and that I would be back later. It was a nice, warm day, and, as I walked through the woods, I could see that it was the ideal place for prime time training.

I went back to the site, got one of the other NCOs in the platoon and the two of us went back into the woods. As we walked, we drew the outline of the woods on paper and then put in some plans for the training. When the platoon leader and I explained the plans to the commander, he rode up with us to look at the new training site. I had explained to the commander that it was a good place to train, because the soldiers would be away from the tac-site. They wouldn't be able to go back to their section, and they wouldn't get bored, because the training would be realistic.

The way we had it set up was, one sergeant would march down the road with about ten soldiers who were dressed for war, to include their weapon and mask. As they walked down the road, I had an ambush set up, including a .50 caliber machine gun. When the shooting started, the soldiers would hit the ground and return fire and the leader would take charge and have the soldiers overtake the bunker.

Farther down the road there was a gas attack, and when that was over, we had about five stations where the soldiers could train. We even had the mess hall bring chow out to the woods. The training went so well, that the whole unit had to go through it. One day, the commander had the battalion commander come down to check out the training, and before you knew it, the whole battalion was doing the same type of training. If there were no woods available to be used for training, they had to go into a field to do it.

There were also times when the unit had to wear MOPP4 (Mission Oriented Protective Posture—Stage 4), for an hour or more. When that would happen, I would take the soldiers in the platoon into the woods, and we would walk around in MOPP4. I told them it would do no good to sit around in the gear, because they may have to move around in it, should they go to war.

Challenging and worthwhile training both creates and reflects unit cohesion. The soldier gains confidence in himself, his fellow soldiers, and his leaders, as well as personal competence and confidence in his weapons and equipment, through successful completion of challenging training. Soldiers need to know that, as a cohesive team, they can carry the fight to the enemy and win.

An important aspect of training for combat is to help the soldier learn how to deal with fear. The leader can first teach the soldier about the

KNOWING AND TRAINING YOUR SOLDIERS 45

physical effects of fear and he can develop training tasks that require moral and physical courage. The soldier should face situations in training that generate fear and anxiety, so that he can learn to deal with them. The leader should also tell soldiers that extreme fear occurs in combat, he must prepare for it in advance, and that fear is greatest:

- Just before the action
- When in defense and under artillery attack
- Under bombing attack at night
- When uninformed about the situation and helpless to retaliate

Knowing what symptoms of fear are and when to expect them makes the soldier's situation more predictable. All men feel fear in combat; it is a normal human response.

Soldiers should also practice, in training, the types of communications required in combat. The new, inexperienced soldier may find himself alone, in a hostile and dangerous environment, and out of contact with those who directed his movement in training. The soldier must be aware that loss of communication may occur, he must be taught what to do when it happens, and given a chance to react to it in a field exercise.

Leaders must be soldiers who have shown a willingness to assume responsibility in training. A crucial task of the leader is to instill and develop pride and spirit in the unit, by building a sense of personal responsibility through assigning responsibility and holding the soldier accountable for their actions. Pride comes from respect for the unit's ability. The measurement of successful training should be the capability to meet an attainable, realistic standard, rather than just completing a block of training hours.

A soldier who lacks pride in himself and his own performance, feels no pride in his unit or his leaders. It's important, therefore, that the leaders show respect for each soldier to encourage self-esteem and pride, so that the soldier can have a sense of pride in his unit.

TRAINING THE TEAM DURING COMBAT

The training stage during combat is affected by:

- The dimensions of time and space
- The felling of soldiers

- The level of critical information
- The environment

In terms of time and space, the team literally develops under fire. In peacetime, the unit has time to practice training mission. The leader must use any available time to sharpen basic combat skills. In combat, the soldier will have to learn what the realistic threat is, how the enemy thinks and operates, and how to react to the real situation in response to enemy movement and activity.

On the battlefield, time is critical; soldiers lives are at stake. The most effective leaders will realize that team building can, and, must be made to work in any environment if the leader follows some basic principles:

- Know the job
- Know the soldier
- Develop the soldier
- Structure the situation for the soldier

The primary concern of most soldiers is the leaders' competence—"Does he know what he is doing?" It's the responsibility of the leader to know the tasks required of his level of rank and experience, as well as the tasks of his soldiers. As the leader gets to know the soldiers, he determines their reliability. He gives them responsibility, where possible, to develop them into potential leaders. He also encourages those few soldiers who do not seem to fit in, by pointing out that being an effective team member is important to their survival and to the survival of the unit in critical war situations.

In combat, the soldier's job expectations will be strongly influenced by his need to survive. On the basis of his own experience, the leader considers the time it takes to get used to the combat environment, and gives the soldier time in which to develop. Guiding this process is the responsibility of the leader.

When in contact with the enemy, the soldier's greatest need is the feeling of structure that his team members and leader provide. The leader structures the situation by ensuring that the soldiers are adequately informed. The soldier wants to know all he can about his situation. The leader must make his presence known, by moving among his soldiers, issuing verbal instructions, using arm and hand signals, using flares, or simply standing up and leading his soldiers, when appropriate.

The leader and the soldiers must be constantly aware that suppressing fearful behavior during combat is critical, because it can spread from soldier to soldier and paralyze an entire unit. During the training stage, the leader must:

- Trust and encourage trust
- Allow growth while keeping control
- Identify and channel emerging leaders
- Establish clear lines of authority
- Develop soldier's and unit goals
- Be fair and give responsibility
- Demonstrate competence
- Know the soldiers
- Pace soldiers' battlefield integration
- Provide stable unit climate
- Develop safety awareness for improved readiness

SUSTAINING THE COMBAT-READY TEAM

It's the leader's responsibility to sustain team spirit and effectiveness once a cohesive team develops. The sustainment stage is characterized by accomplishing the mission through teamwork and cohesion.

It begins when the soldier and leader emerge from the questioning and challenging stage and begin to work together, as a team. During this stage, the team, rather than individuals, accomplishes tasks and missions. The leader must listen to what the soldiers say, how it is said and what the soldier does not say. The leader knows that listening and acting to improve the situation are powerful means of gaining trust and developing cohesion. The good leader is always alert to suggestions, complaints and input from soldiers.

As soldiers develop their personal skills and blend them into team training, they become more and more proficient as a team Realistic training can be conducted as the leader analyzes the risks involved and integrates safety considerations into the training scenario. Aside from relieving boredom and developing teamwork, demanding team training enables soldiers and leaders to feel more capable to do their job in combat.

Maintenance is essential to sustaining the fighting spirit of combat-ready teams. In such teams, soldiers develop special relationships with their weapons and equipment—at times they even give them names. If the unit goes into combat, there will be no time to stand down for maintenance. The leader must also do all, in his power, to ensure timely delivery of supplies to his team. This reduces the fear of isolation that soldiers might feel.

To sustain his team, the leader must demonstrate caring leadership through his entire time in the unit. If military necessity dictates some hardship for the soldier, the leader must show an understanding attitude, and then communicate, precisely, why the soldier cannot be allowed to do all that he might want to satisfy his personal concerns. Unit activities provide a focus around which members come together and create an atmosphere for emerging relationship and unit cohesion. Care must be taken to avoid overemphasizing unit activities, because it can be damaging if it takes the focus of the unit away from mission accomplishment.

Participation in military ceremonies fosters pride and spirit in the unit and in the Army. Such unit spirit is essential in building cohesive teamwork. A unit sports program can give all the soldiers a sense of membership in the unit, they begin to refer to the company team as "our" softball, volleyball or track team. When they do so, they identify with their unit. Soldiers recall and talk about highlights of competition, reinforcing mutual feelings and building cohesion. The variety of social activities is limited only by time, imagination of the planner and good taste.

The unit party provides a relaxed atmosphere for soldiers to develop positive relationships among themselves and with their leader. It also provides an opportunity for families to meet other families and enhances family belonging to and involvement in the unit.

Encouraging soldiers to develop their spiritual lives is another way in which the leader can influence the cohesion of the unit. Through encouraging his soldiers to practice and develop their faith, the leader shows another facet of his concern for their well-being. The unit chaplain can assist in answering any questions the leader may have in this area.

SUSTAINING THE TEAM DURING COMBAT

Conditions in combat exact pressure on the leader's efforts to sustain his team. He must know how to deal with each situation. Conditions that undermine teamwork are; continuous operations, enemy operations, rumors, boredom and casualties. The continuous operations anticipated on the modern battlefield cause effects such as:

KNOWING AND TRAINING YOUR SOLDIERS 49

- Decreased vigilance
- Reduced attention
- Slowed perception
- Inability to concentrate
- Mood changes
- Communication difficulties
- Inability to accomplish manual tasks

If left unchecked, these effects can deteriorate the most cohesive teams and damage their will to fight. Proper sleep and rest are necessary to keep soldiers functioning at their best. The leader must get rest and sleep, also.

The appearance of the enemy in force, or fire from an enemy that cannot be seen can affect the soldier's performance as a team member. The more he knows about what to expect and how to react, the more confident he will be in the moment of crisis. During breaks in combat, the team should spend time discussing recent combat actions, their performance, and ways that they can improve.

Casualties create personnel turbulence and have a psychological effect on the soldier. Proper safety precautions can assist in minimizing unnecessary casualties and their psychological impact on the soldier. The soldiers must have no doubt that if they are injured, they will not be deserted because of hostile fire. The loss of a leader, because of injury or death, will even more seriously affect teamwork. When a new leader is appointed, other leaders need to back and support him, even if he has no combat experience, he still has to fit into the unit. Each new leader has to depend on the soldiers and on other leaders to assist him in adapting his training and peacetime experience to combat.

Remember the Captain I was telling you about, the one that was overweight? Well, like I said before, he was a very good commander. One day he gave me a sheet of paper with a list of names on it, to include my name. That list had the name of the leaders in the unit that would take command, should he be killed in combat.

When I saw that I was number eleven on the list, it did something to me, because I had never thought of doing anything but being a platoon sergeant, at the time. It had never occurred to me that something could happen to the Commander, XO, and other officers and NCOs in the unit, but, because most of the time they were all together, I supposed it very possibly could. Anyway, after looking at the list, I started learning more about running the unit control center. He was the only commander in my twenty-one years of service that ever had a list of who would be in charge. I think that's one of the reasons why he was the best commander I had when I was on active duty.

Dealing with boredom is essential for combat effectiveness. Effective leaders focus on security, resupply, personal hygiene, patrol activities, cross training and radio procedures to keep their soldiers from getting bored.

Rumors are bits of information that are not based on definite knowledge. To sustain teamwork, the leader must constantly use truth to deal with rumors and put them to rest. The leader can help control rumors by stressing honesty, keeping the soldiers informed about what is going on, and clear up all rumors with facts, as soon as possible. He should also identify and counsel those who spread rumors, but, at the same time, be careful to avoid wrongly accusing team members of starting rumors, as this creates distrust.

It's important that leaders learn how to deal with soldiers feelings. When a soldier is threatened, he may feel anger, despair or fear. If he is feeling angry, that could indicate a high level of confidence. The leader's challenge is to direct that soldier's anger in the right direction. If a soldier is afraid, it simply indicates that the soldier may or may not take action to eliminate the threat. It may depend on how the soldier can deal with fear, or how skillful the leader is in controlling the undesirable effect of the fear on himself and his soldiers.

Fear can come from many directions in combat and it can immobilize soldiers, destroy a team's will to fight, and lead to despair and panic. If the leader can reduce fear levels, he can inspire effective action. He can, and must, emphasize that fear is normal. Fear control is a central function of combat leadership—controlling fear can prevent panic.

Soldiers in panic have intense fear, are easily spooked, and tend to flee the battlefield. When these conditions exist, a "trigger" incident confirms the belief that the situation is out of control and can cause the soldier to panic. If a soldier turns and runs, others may follow and the action may snowball until the entire unit is in flight. To prevent panic, the leader must focus on and control what the soldiers believe to be true.

Soldiers in combat are subject to all the fears that lead to panic. When trigger incidents occur, the leader must follow with prompt and calm actions. He can:

- Keep the soldiers busy with routine and meaningful tasks
- Move from position to position, adding structure to the situation
- Slow the soldiers down so that they can act instead of react

- Set a personal example of fearlessness and insist that all on the leadership team do the same
- Explain reason for withdrawals and delaying actions
- Stress the unit's ability, as a unit, to cope with all battlefield situations
- Assure the unit that it is in command of the situation and not in an inescapable one
- Assure the unit that it's flanks, rear and supplies are secure, if this is the case

If panic develops in spite of all the leader's efforts, he must take firm and decisive action to stop it as soon as possible. Once the panic is stopped, the leader must immediately restructure the situation and give the panicked soldiers something constructive to do as part of the larger unit. The leader must work constantly to restructure the situation and keep the unit organized. Some of the actions he can take to help restructure some of the situations are:

- Use chain-of-command to avoid conflicting orders and prevent rumors
- Manage time efficiently to prevent prolonged waits
- Avoid false alarms
- Train junior leaders to take command in the event of death of their leaders
- Prevent surprise by stressing security
- Keep the soldiers informed on all matters
- Never express dissention in the presence of soldiers
- Forcefully correct those soldiers who are increasing fear by irresponsible talk

The leader must act to overcome detrimental effects of combat, such as conditions that lead to fear and panic. During the sustainment stage, the leader must:

- Demonstrate trust
- Focus on teamwork, training and maintenance
- Respond to soldiers problems
- Develop more challenging training
- Build pride and spirit through unit military, sports, social and spirit activities
- Observe sleep discipline

- Sustain safety awareness
- Inform soldiers
- Know and deal with soldiers' perceptions
- Keep soldiers productively busy
- Use after actions reviews
- Act decisively in the face of panic

3

TEAM TRAINING WITH HIGHER HEADQUARTERS

PRINCIPLES OF TRAINING

A good leader is one who trains his soldiers well. Putting the soldier first means making sure he is trained to succeed in doing his job.

Like the principles of leadership, the principles of training are guidelines that provide the cornerstone for training. The principles are:

- Train as combined arms and service teams
- Train as you will fight
- Use appropriate doctrine
- Use performance-oriented training
- Train to challenge
- Train using multi-echelon techniques
- Train to maintain
- Make commanders the primary trainers

Today's Army doctrine requires combined arms and services teamwork. When committed to battle, each unit must be prepared to execute combined arms and service operations, without additional training or lengthy adjustment periods. Combined arms proficiency develops when teams train together. The full integration of the combined arms team is attained through the "slice" approach to training management. This approach acknowledges that the maneuver commander controls service support systems.

The goal of combat-level training is to achieve combat-level standards. In order to accomplish that, you and your soldiers must train as you will fight. Every effort must be made to attain this difficult goal. Within the confines of safety and common-sense, leaders must integrate such realistic conditions as smoke, noise, simulated NBC, battlefield debris, loss of key leaders, and cold weather. They must seize every opportunity to move soldiers out of the classroom into the field where they can fire weapons, maneuver as a combined arms team, and include joint and combined operations. He must use appropriate doctrine for his training, and it must conform to the Army doctrine.

Soldiers, who have just been assigned to units, will have little time to learn nonstandardized procedures. Units, therefore, must train on peacetime training tasks to the Army standard, contained in mission training plans, battle drill books, soldiers manuals, regulations and other training and doctrinal publications. Use performance oriented training (hands-on), because soldiers learn best by doing. Leaders are responsible to plan training that will provide these opportunities.

All training assets and resources, to include simulators, simulations and training devices, must be included in the strategy units. Leaders and their soldiers must train to challenge tough, realistic and intellectually and physically challenging training. This type of training both excites and motivates soldiers and leaders. It builds competence and confidence by developing and honing skills.

Challenging training inspires excellence by fostering initiative, enthusiasm and eagerness to learn. Once individuals and units have trained to a required level of proficiency, leaders must structure collective and individual training plans to repeat critical task training at a minimum frequency necessary for sustainment. Leaders must train to maintain proficiency, for their soldiers and for themselves.

Sustainment training is often misunderstood, although it is a reasonable common-sense approach to training. Sustainment training must sustain skills to high standards often enough to prevent skill decay, and to train new people. Army units cannot rely on " peaking" to the appropriate level of wartime proficiency. This is exactly what many units, active duty and reserves, did at the start of the Gulf war. Sustainment training enables units to operate in a " band of excellence", by repetitive critical task training during prime training periods.

Training, using multi-echelon techniques, is the most efficient way of training and sustaining a diverse number of mission essential tasks, within limited periods of training time. To use available time and resources most effectively, Commanders must simultaneously train individuals, leaders and units, at each echelon in the organization, during training events.

Leaders, as well as the soldiers, must also train to maintain all assigned equipment in a high state of readiness, in support of training or combat employment. Maintenance is a vital part of every training program. Maintenance training, designed to keep equipment in the fight, is as important as the soldiers' expert level of proficiency while employing the equipment.

Making commanders the primary trainers is ideal, because the leaders in the chain-of-command are responsible for the training and performance of their soldiers and units, and are the primary training managers and trainers for their organizations. To accomplish their training responsibilities, commanders must:

- Base training on wartime mission requirements
- Identify applicable Army standards
- Assess current levels of proficiency
- Provide the required resources
- Develop and execute training plans that result in proficient individuals, leaders and units

MISSION ESSENTIAL TASK LIST (METL) DEVELOPMENT

Commanders must selectively identify the tasks that are essential to accomplishing the organization's wartime mission. There are two primary inputs to METL development; **War Plans**, which are the most critical inputs to METL development, are the organization's wartime operations and contingency plans, and **External Directives**, which are additional sources of training tasks that relate to an organization's wartime mission. Some examples are:

- Mission Training Plans
- Mobilization Plans
- Installation Wartime Transition and Deployment Plans
- Force Integration Plans

In some cases, these directives identify component tasks, which make up the wartime mission. Commanders analyze the applicable tasks contained in external directives and, select for training only those tasks essential to accomplish their organization's wartime mission. This selection process reduces the number of tasks on which the organization must train.

To provide battle focus on the most important wartime requirements, the Commander identifies specified and implied mission essential tasks, contained in external directives. The following fundamentals apply to the METL development:

- The METL is derived from the organization's wartime mission and related tasks in external directives.
- Mission essential tasks must apply to the entire organization.
- Each organization's METL must support and complement their higher headquarter's METL.
- The availability of resources does not affect the METL development.
- The seven battlefield operating systems (BOS) are used to systematically ensure that all elements of the organization's combat power are directed toward accomplishing the overall mission. The systems are as follows; maneuver, fire support, support, command and control, intelligence, mobility/survivability and air defense.

In similar type organizations, mission essential tasks may vary significantly, because of different wartime missions or geographical locations. Battle books can also assist in the identification of METL. They contain detailed information concerning warplans, such as tactical routs to wartime areas of operation, ammunition upload procedures, execution of schemes of maneuver and other support requirements.

After the Commander designates the collective METL required to accomplish his organization wartime mission, the CSM and other NCOs develop a supporting individual task list for each mission essential task. After the mission essential tasks have been identified, the Commanders establish supporting standards and conditions for each task.

PLANNING FOR THE MISSION

Planning is an extension of the battle focus concept that combines organizational METL with the subsequent execution and evaluation of training. The Commander provides the principal inputs at the start of the planning process, the METL and the training assessment. The training assessment of each separate METL enables the Commander to develop his training vision. His training vision is supported by organizational goals that provide a common direction for all of the Commander's programs and systems.

There are three kinds of training plans, long-range, short-range and near-term plans. Properly developed training plans will:

- Maintain a consistent battle focus.
- Be coordinated between associated combat, combat support and combat service support organizations.
- Focus on the correct time horizon.
- Be concerned with future proficiency.
- Cause organizational stability.
- Make the most efficient use of resources.

The command training guidance (CTG) is published at division and brigade levels to document the organization's long-term training plans. It will be used as a ready reference for the planning, execution and assessment of training throughout the long-range planning period. Examples of topics normally addressed in the command training guidance are:

- METL and associated battle tasks.
- Commander's training philosophy
- Combined arms training
- Major training events and exercises
- Leaders' training
- Individual training
- Mandatory training
- Standardization
- Training evaluations and feedback
- New equipment training and other force integration considerations
- Training management

All echelons, from division to battalion, publish the long-range planning calendar concurrently with the CTG. The long-range planning calendar will normally extend at least two years into the future. During long-range planning, commanders organize training time to support mission essential training and concentrate training distractors in support periods.

During long-range planning, commanders and their staffs make a broad assessment of the number, type and duration of training events required to accomplish METL training. Major training events are the common building blocks that support an integrated set of METL-related training requirements. During planning, senior commanders allocate maximum training time to subordinates.

Short-range training plans define, in greater detail, the broad guidance on training events and other activities contained in the long-range training guidance and long-range calendar. Each level, from division through battalion, publishes short-range training guidance that enables the Commander and staff to prioritize and refine mission essential training guidance contained in the long-range CTG. Commanders must publish the short-range training guidance and will allow sufficient time to ensure subordinate units have the time to develop their own short-range plans.

The division provides quarterly training guidance to subordinate commands and installations at least ninety days prior to the start of each quarter. An important aspect of the QTP is the role of the NCOs. Within the framework of the Commander's guidance, the CSM and other key NCOs provide planning recommendations on the organizations' individual training programs. They identify the individual training tasks that must be integrated into the collective mission essential tasks, during the short-range planning period. Examples of topics, normally addressed in the quarterly training guidance, are:

- Commander's assessment of METL proficiency
- Combined arms and service training
- Training priorities
- A cross reference of training events and associated METL training objectives
- Individual training
- Leaders' development
- Preparation of training and evaluations
- Training evaluation and feedback
- Force integration

- Resource guidance
- Training management

The short-range planning calendar refines an applicable portion of the long-range planning calendar. In short-range planning, commanders allocate training resources to subordinate organizations for specific training activities.

SHORT-RANGE PLANNING BRIEFING

The short-range training briefing is a conference, conducted by senior commanders, to review and approve the training plans of subordinate units. Division Commanders receive the short-range training briefing from subordinate brigades and all battalions in the division.

Training briefings produce a training contract between the senior commanders and each subordinate commander. During the training briefing, the subordinate commanders, at a minimum, usually address these specific areas:

- A review of the last short-range planning period's accomplishment and short falls.
- The organization's METL and assessment of proficiency levels
- A discussion of the unit's training focus and objective for its upcoming training period.
- A presentation of the organization's short-range planningc calendar.
- A description of upcoming training events.
- Leader development program, with emphasis on officer warfighting skill development.
- Approach to be used for preparing trainers and evaluators.
- Force integration plans for the upcoming period.
- Resource allocation.

The CSM provides an analysis of the unit's individual training proficiency and discusses the unit's planned individual training and education. The following topics are examples of this briefing:

- Individual training proficiency feedback received, concerning previous short-range planning period.
- An assessment of the organization's current individual training proficiency.

- Individual training events planned during the upcoming short-range planning period and strategy to prepare soldiers for these evaluations.
- A description of METL-derived individual tasks to be integrated with upcoming collective mission essential tasks.
- Marksmanship and physical fitness programs.
- The organization's education program.
- The NCO leader development program and its relationship to improving warfighting skills.

NEAR-TERM PLANNING

Near-term planning is primarily conducted at battalion and subordinate command levels. It is conducted to:

- Schedule and execute training objectives specified in the short-range training plan to Army standards.
- Make final coordination for the allocation of resources to be used in training.
- Provide specific guidance to trainers.
- Complete final coordination with other units that will participate in training as part of the combined arms or service slice.
- Prepare detailed training schedules.

Near-term planning covers a six-to-eight week period, prior to the training being conducted. Formal near-term planning culminates when the unit publishes its training schedule. Near-term planning includes the conduct of training meetings to create a bottom-up flow of information regarding specific training proficiency needs of the small unit and individual soldier.

At battalion level, training meetings primarily cover training management issues; at company and platoon level, they are directly concerned with the specifics of conducting training. Training schedule formats may vary among organizations, but they will :

- Specify when training starts and where it takes place.
- Allocate the correct amount of time for scheduled training and also additional training, as required, to correct anticipated deficiencies.
- Specify individual, leader and collective tasks to be trained.

TEAM TRAINING WITH HIGHER HEADQUARTERS

- Provide concurrent training topics that will efficiently use available training time.
- Specify who conducts the training and who will evaluate the results.
- Provide administrative information concerning uniform, weapons, equipment, references and safety precautions.

Training is locked in when the training schedules are published. Command responsibility is established as follows:

- The Company Commander drafts the training schedule.
- The Battalion Commander approves the schedule and provides necessary administrative support.
- The Brigade Commander reviews each training schedule published in his command.
- The Division Commander reviews selected training schedules in detail, and the complete list of organization wide training highlights, developed by the division staff.

Senior leaders provide feedback to subordinates on training schedule quality, and subsequently attend as much training as possible, to ensure that METL are accomplished to standards.

EXECUTION AND ASSESSMENT

Decentralization tailors training execution to available resources and promotes bottom-up communication of unique wartime mission-related strengths and weaknesses of each individual, leader and unit. Senior leaders must personally observe and evaluate the execution of training at all echelons.

By allotting quality time for personal visits to training, senior leaders communicate to the entire chain-of-command that training is the organization's top peacetime priority. They also observe and assess the quality of training at all echelons, down to the lowest levels of the organization. They receive feedback from subordinate leaders and soldiers during training visits.

The most beneficial senior leader and staff visits to training are unannounced or short notice. The leader, then, observes normal training, as experienced by soldiers, and prevents excessive visitor preparation by subordinate organizations. All good training, regardless of the specific collective and individual tasks being executed, must comply with certain common requirements. The requirements are adequate preparation, effective presentation and practice, and thorough evaluation.

Informal planning and detailed coordination, known as pre-execution checks, continue until the training is performed. Pre-execution checks cover the preparation of the individuals to be trained and the training support required. Properly prepared individuals are trained on prerequisite tasks prior to training. Trainers are coached on how to train, given time to prepare, and the training rehearsed so that it will be challenging and doctrinally correct. The trainers may use any combination of demonstrations, conferences, discussions and practice activities to present training.

If individuals or organizations are receiving initial training on a mission essential task, trainers emphasize on the basic conditions. Properly presented and practiced training is accurate, well structured, efficient, realistic, safe and effective.

Although individuals and organizations may sometimes compete against one another, they should always compete to achieve the prescribed standard. Some of the considerations for conducting effective training are:

- Battle rosters
- NCO training responsibilities
- Training and evaluation outlines
- Staff training
- Leader training

EVALUATION OF TRAINING

Evaluation of training measures demonstrates the ability of individuals, leaders and units against specified training standards. Evaluation can be formal, informal and external. Evaluation for individual and small unit training, normally includes every soldier and leader involved in the training.

During and after the evaluation, evaluators prepare their findings and recommendations. They provide these reports to the evaluated unit commander and higher commanders, as required by the headquarters directing the evaluation. The after-action review provides feedback for all training and allows training participants to discover, for themselves, what happened, why it happened and how it can be done better. The after-action review consists of four parts, which are:

1. Establish what happened
2. Determine what was right or wrong with what happened
3. Determine how the task should be done differently the next time
4. Perform the task again

Following the after-action review with all participants, senior trainers may use the AAR for an extended professional discussion. This usually includes a more specific AAR of leader contributions to the observed training results. Evaluators must be trained as facilitators to conduct after-action reviews that elicit maximum participation from those being trained. Experience has shown that providing qualified individuals to evaluate others is well justified. Senior leaders ensure that evaluations take place at each level in the organization. They also take advantage of evaluation information to develop appropriate lessons learned, for distribution throughout their command.

Senior leaders use evaluation information as one component of the feedback system. Some sources of training feedback available to senior leaders are:

- Training planning assessment
- Senior, lateral and subordinate headquarters training plans
- Quarterly training briefings
- Yearly training briefings
- Resource allocation forums, such as PBACs or range scheduling conferences
- Personal observations
- Leader development discussions
- Staff visits
- Evaluation data

Evaluation reports provide the chain-of-commander with feedback on the demonstrated training proficiency of individuals, leaders and units, relating to specific training events and objectives. Assessments are neither limited to the training cycle, nor strictly related to training issues. The feedback that occurs during organizational assessment, allow synchronization of all functions and echelons of an organization.

4

WHO WE MUST FIGHT, AND WHY

NEW CONCEPTS IN ARMY DOCTRINE

The mission of the U. S. Army is to protect and defend the Constitution of the United States of America. As the sole remaining superpower to emerge from the cold war ear, the global strategic view of the United States has turned toward new and multiple regional concerns.

The Army must be able to react faster than the enemy, which will enable it to seize and hold the initiative. This means that leaders must anticipate events on the battlefield and perform such activities as; jamming enemy communications, suppressing enemy air defenses, shifting reserves, securing decisive terrain, depriving the enemy of resources, gaining information, holding the enemy in position, disrupting his attack, and setting up for future operations.

Commanders must think about the use of all available target acquisition and attack assets to operate throughout the depth of the battlefield. The aim is to apply combat power, simultaneously, throughout the depth and space of the battlefield, stun the enemy and then defeat him. This method of operation was used during the beginnings of Operation Just Cause and Desert Storm.

Commanders will have to visualize the battle space in which they will fight, under more various and ambiguous conditions than before, know the consequences of decisions and when to make them, and anticipate the future course of events.

THE ARMY OBJECTIVES

Every soldier should read FM 100-1, because it provides the historical and legal foundation of the Army. He will learn that the basic objective of the Army is to defeat the enemy's forces on land. During peacetime, the mission of the Army is to deter war and prepare for the wartime mission. The Army's mission, during war, is to render the enemy's forces ineffective.

There is the "general war", which is an armed conflict between major powers, in which the total resource of the armed forces may be in jeopardy. On the other hand, a "limited war" is an armed conflict between two or more nations, at an intensity below that of a general war, where means and/or ends are constrained. Then there is the "low intensity conflict", which is a limited politico-military struggle to achieve political, social, economic or psychological objectives.

CONSTITUTION

When you join the Army, you raise your right hand and make a pledge to support and defend the Constitution of the United States. The legal basic and framework for a military establishment, charged to provide for the common defense, is found in the preamble to the Constitution of the United States. Title 10, United States Code, Section 3062, defines the primary roles of the Army.

The Constitution states that **only** congress would have the power to raise and support armies and declare war. It gives the President the power (as the nation's chief executive) to command the armed forces as the Commander-in-Chief. The Constitution states that only congress has the power to declare war. This provides a debate, by many men, to determine the worthiness of the cause, prior to American troops putting their lives on the line. It also keeps one person (President) from having the power, alone, to sacrifice the lives of many for what he thinks is right or wrong.

NATIONAL SECURITY

The National Security Act of 1947 established the current structure for national defense. The National Security Act (as amended) places the Departments of the Army, Navy, Air Force and Coast Guard under the direction, authority and control of the Secretary of Defense. The basic national security policy is to preserve the United States as a free nation, with our fundamental institutions and values intact. The broad objectives of our national security policies are:

- Deterrence of any kind of attacks against the United States, its forces and allies, and discouragement of the employment of nuclear weapons.
- Deterrence of any kind of attacks against the United States, our allies, and vital United States interests worldwide, including sources of essential materials, energy and associated lines of communication.
- Encouragement and assistance to other nations in defending themselves against armed invasion, insurgencies and terrorism.
- Discouraging the enemy from attempting coercion of the United States, its allies and friends.
- If deterrence fails, fight at the level of intensity and for the duration necessary to attain United States' political objectives.

UNITED STATES POLICIES ON NBC

The United States policy on nuclear warfare, is to deter it by means of a strong United States nuclear warfare capability. In other words, the more we have, the less likely other countries are likely to use it against us.

We are prohibited from using biological weapons, **period**.

As for chemical weapons, the policy states that U. S. troops cannot use chemical weapons **first** in a war.

When I went to war in Vietnam, I didn't know any more than the soldier that went to the Gulf, did. I do know that many soldiers died in Vietnam because there were many untrained leaders. After the war, leadership schools became a big issue, and it's still that way, today.

You must take care of your soldiers and get them back, alive, from the combat zone to their loved ones. There are many leaders, today, that still can't sleep at night from thinking about their soldiers that were killed in action, not because the enemy was better, but, because they didn't train their soldiers to survive. Take care of your soldiers, and they will take care of you. Then, if you should ever go to war, you and your soldiers will return to a country that's depending on you.

PART II

GETTING PROMOTED

5

JUNIOR NCO'S PROMOTION (SGT AND SSG)

SEMICENTRALIZED PROMOTION

Field grade commanders in the grade of LTC or higher have promotion authority over sergeants and staff sergeants. Sergeant and staff sergeant promotions are called semicentralized promotions because board appearance, promotion points calculation, promotion list maintenance and the final execution of the promotions occur in the field in a decentralized manner. Promotion point cutoff scores are determined and announced monthly for each MOS by HQDA (centralized). The unit's PAC will prepare the promotion certificates and they will be signed by the promotion authority or a higher level commander.

KEY EVENTS AND WORK SCHEDULING

Listed below is the complete semicentralized promotion cycle.

Key Events

- During the month of the promotion board and the month before the battalion S-1 (BNS1) will identify eligible soldiers and prepare section A of the promotion point worksheets (DA Form 3355), obtain approval and forward them to the PSC.

- During the board month the PSC will prepare section B of the promotion point worksheets and return them to the battalion. The promotion board will be conducted, approval obtained and forwarded to the PSC. The soldiers promotion points will be recorded in the SIDPERS database.

- Board month plus two the promotion work center will receive the promotion point cutoff scores, identify promotees, verify eligibility and issue orders.

Work Scheduling

Every month a semicentralized cycle starts for a new group of soldiers and this cycle must work between PSC and the Battalions in order to avoid hang-ups in the promotion system. At the battalion level, between the 1st and the 10th of the month, new recommendations for the next month's board will be prepared. The unit commander's decision for the promotion will be obtained along with the promotion authority's signature on DA Form 3355. The DA Form 3355s will then be forwarded to the PSC.

Between the 1st and the 15th of the month, the battalions will prepare for, conduct and finalize the current month's board. The promotion authority's signature will be obtained on DA Form 3355 and the promotions packets from the current month's board will be forwarded to the PSC. At the PSC level, section B of the DA Form 3355 will be prepared and the form will be returned to the BNS1.

Between the 11th and 20th of the month, the PSC will obtain the cutoff scores, identify eligibles and verify eligibility. Orders will be issued between the 18th and 25th of the month. Between the 18th and the 30th, the PSC will submit SIDPERS transactions and record the current month's board results.

THE PROMOTION PACKET

The promotion packet consists of the following:

- Promotion point worksheet (DA Form 3355) is used for initial board appearance.
- Approved report of the promotion board proceedings.
- Latest (current or previous) two DA Form 3355s used for recomputation or reevaluation.
- A copy of the soldier's latest AAC-C10 report.
- A copy of any document that allows the soldier's previously determined promotion score to be adjusted.
- A copy of any document used to confirm points of DA Form 3355 that is not filed in the MPRJ under the provisions of AR 640-10. All DA Form 3355s that are no longer required for filing in the MPRJ will be removed and given to the soldier.

BECOMING ELIGIBLE FOR PROMOTION

The BNS1 will prepare a list of personnel eligible for SGT/SSG promotion consideration or receive C01 reports from the PSC. They will then print out the report and forward it to the appropriate unit. When the report gets to the commander, he or she will personally review it and identify soldiers to be recommended for promotion by annotating yes on the form if the soldier is recommended or no if not. After the commander has signed the form, it will be returned to the BNS1. The BNS1 will prepare Section A of the DA Form 3355, obtain approval and forward it to the PSC.

If a soldier is eligible for promotion without a waiver, but is not recommended, he or she will be counseled by the commander as to why no recommendation was made, and what they can do to correct deficiencies. The statement of counseling at the end of the DA Form 3355 will be completed by the soldier.

RECOMMENDATION FOR PROMOTION

Before getting promoted to sergeant or staff sergeant, more things have to happen other than the BNS1 forwarding a DA Form 3355 to the unit for the commander's approval. Most likely the first line supervisor will recommend the soldier for promotion. This recommendation may or may not be in writing.

Next, the commander may talk to the platoon or section leader along with the platoon or section sergeant about the soldier. The first sergeant and the XO may also be asked what they think about the soldier's recommendation for promotion. After talking with the other leaders, the commander will decide whether or not to recommend the soldier for promotion.

ELIGIBILITY CRITERIA

Before the commander can recommend a soldier for promotion, the soldier must be eligible to compete for and be promoted in his or her primary MOS. They must have a high school diploma or equivalent (GED) or an associate or higher degree. The soldier must be a graduate of PLDC prior to competing for promotion to staff sergeant. The soldier must appear before a promotion board and be recommended by the board voting members. The soldier must have no record of a court martial conviction and must not be ineligible to reenlist in accordance with AR 601-280. They must be physically qualified and must pass the Army Physical Fitness Test and must be qualified on their individual weapon.

Soldiers competing for promotion to SSG must obtain 550 points and those competing for SGT must obtain 450 points prior to being added to the recommended promotion list. Sergeants must have 12 months of service remaining on active duty after being promoted to staff sergeant. In order to meet this requirement, they may extend the required number of months needed, as long as the extension does not go over the 20 years a SSG can stay on active duty.

TIME IN GRADE AND TIME IN SERVICE

The promotion authority or designee may waive the eligibility requirement of TIMIG or TIS. For promotion to SSG in the primary zone, a SGT will need 84 months TIS, but can go before the promotion board with 81 months. To compete in the secondary zone, he or she will need 48 months TIS but can appear before the promotion board with only 45 months. Before appearing before the board for SSG, they will need 10 months time in grade for the primary and 5 months for the secondary zone.

For promotion to SGT in the primary zone, a SPC or CPL will need 36 months TIS, but can appear before the board with 33 months. To compete in the secondary zone, they will need only 18 months, but can be boarded with 15 months. The TIMIG requirement is 8 months for promotion in the primary zone, but can be waivered to 4 months. All times are as of the first day of the promotion month.

The secondary zone provides incentives to those who strive for excellence and whose accomplishments, demonstrated capability for leadership and marked potential warrant promotion ahead of their peers.

RULE FOR CONDUCTING THE PROMOTION BOARD

The sergeant and staff sergeant promotion board will be conducted by the 15th of each month. The board voting members will use a question and answer format, only. Soldiers will not be required to perform hands-on tasks. Soldiers going before the promotion board must be fully qualified in their PMOS. The promotion or convening authority will appoint, in writing, at least three voting members and a recorder without vote. The voting members may be all officers, all NCOs or mixed.

The president of the board will be the senior member (preferably a field grade officer) or, for an all NCO board, a CSM or SGM if a CSM is not available. Board members will be senior in grade to those being considered for promotion and unbiased. At least one voting member will be of the same sex as the soldiers being evaluated. If this is not possible, the reason will be recorded as part of the board proceeding. Board members will be composed of an ethnic mixture, even though the board may not be considering soldiers of minority ethnic groups. The board should not be composed of minority ethnic group members exclusively. A nonvoting recorder should be from the BNS1, but need not be senior in grade to those going before the promotion board.

If a great number of soldiers are appearing before the promotion board, it may be split into two or more panels provided each panel consists of at least three voting members and a recorder without vote. To expedite the process, each soldier should appear before only one panel.

The board members may also be tasked to consider soldiers for removal from a recommended list.

Once a board is convened, the same members will be present during the entire board proceedings. The board president may choose to be a voting member or to vote only to break a tie. Each member will have one vote, and will complete a DA Form 3356 (Board Member appraisal worksheet) to vote on each soldier.

Use of the MPRJ by board members is optional. The recorder will prepare the DA Form 3357 (Board Recommendation) after each soldier appears and obtain the board president's signature. The recorder will also complete the remaining portion of section C of the DA Form 3355.

The promotion authority will brief the board president on his or her responsibilities. If a tie exists, the board president will vote to break the tie. The BNS1 will notify the unit commanders of the board schedule and prepare the record of the board proceedings. The unit will notify the soldier as to the date and place of the board. The board president will review the report of the board proceeding for accuracy, sign the report and forward it to the promotion authority for approval or disapproval.

Within 3 working days after the promotion board adjourns, the promotion authority will approve or disapprove the report in its entirety. If the promotion authority cannot accomplish these actions within 3 working days after the board adjourns, a memorandum of explanations, signed by the promotion authority will be attached to the DA Form 3355 citing the specific reason for the delay. Disapproval cannot be used to disagree with the board's recommendation. The approval or disapproval should pertain only to the correct constitution and conduct of the board.

Completion of section C of DA Form 3355 is required and will be accomplished by the promotion authority of the board's organization if approved. If not, each soldier will be advised of the reason, even if they have departed the organization. This will apply to all soldiers considered by the board, whether or not they were recommended.

Prior to being added to the recommended list, soldiers competing for SSG must obtain a minimum of 550 promotion points and those competing for SGT must obtain 450 promotion points. The completed board action should reach the PSC by the 20th day of the month.

Soldiers that were not recommended or did not obtain enough promotion points to be placed on the promotion list will be counseled by the promotion authority or his or her designated counselor and signatures obtained in section C of DA Form 3355.

THE STUDY GUIDE

In order to get promoted to Sergeant (E5) and Staff Sergeant (E6), the soldier must appear before a promotion board. This is the part of the promotion process that many soldiers find difficult, but only because they don't feel prepared. The study guide can be very helpful in preparing for

the board, and the best study guide is the one you make for yourself.
Most study guides are just question and answers about subjects you may be asked questions about. You can make a study guide about subjects you know you will be asked questions about. When putting your study guide together, remember questions concerning the knowledge of basic soldiering will be tailored to include; land navigation, survival, night operations, inclement weather operations, adverse environment and terrain.

Don't spend too much time on questions that a board member may ask, because each member will score you for only the question he or she asked. You can get up to 45 promotion points for questions you may be asked concerning knowledge of basic soldiering, so, if you are asked two questions by one member and get only one correct you may still get 20 points or more. It is up to each board member as to the score you will receive, but, believe me, they want you to pass the board.

All you have to do is take your time, and if you are not sure of the question, just ask them to repeat it for you. It's not hard to max the board, and most soldiers that fail, do so because they answer a question incorrectly and just give up after that, yet, I've seen many soldiers miss an answer and only lose one point for it.

Each board member will have a DA Form 3356 (Board Member Appraisal Worksheet) and, on the form, there are six categories on which the soldier will be evaluated. They are:

- Personal appearance, bearing and self-confidence
- Oral expression and conversational skills
- Knowledge of world affairs
- Awareness of military programs
- Knowledge of basic soldiering
- Soldier's attitude (includes leadership, potential for advancement and trends in performance)

For each category there is a point spread, depending on whether the board member feels you are an average, above average, excellent or outstanding soldier. The point spread runs from 1-15 points for an average soldier, 11-25 points for an above average soldier, 16-35 points for an excellent soldier and 21-45 points for an outstanding soldier. Each member will, for the purpose of counseling, comment on specific items in which the soldier appears non-competitive and/or weak, in the remarks section of the DA Form 3356. All points from each member's DA Form 3356 will be added for the total score (a maximum 200 points).

If you have never been before a promotion board, try to go before the Soldier of the Month board before the promotion board, because they are conducted in about the same way, and, in some places, they have the same board members as the promotion board. A pre-board is another way to get used to facing board members. If there is not one in your unit, talk to your First Sergeant about planning one for you and the other soldiers.

THE WOMEN'S CLASS A UNIFORM

The Army Green Classic Uniform is authorized for year-round wear. The coat has been designed to look equally well over both slacks and skirt, and that's why it has the longer than average length.

The coat fits snugly around the back of the neck, without a gap or break under the collar. Across the back it fits smoothly, without too much tightness or wrinkles, and with sufficient flexibility for arm movement. It should fit smoothly across the bust without the button at the bust popping open or gapping and easily over the hips and conform to the waistline curve without blousing. The bottom button of the coat should fall about one inch below the natural waistline. The coat front, below the waist, must overlap without pulling or gapping, so the front of the coat will present a straight line from the top button of the coat hem and lie smoothly and flat without twists, pleats or bulges. The back vent must also overlap, without pulling or gapping.

The coat is designed so that, if the coat size is correct in other areas, the coat length, with a few exceptions, will be correctly proportioned. The bottom edge of the sleeve should fall one inch below the bottom edge of the wrist bone and cover the shirt sleeve.

The skirt should fit smoothly over the hips, so that it does not drape in folds. The waistband should fit with one half inch of looseness on either side, between the natural waistline and the waistband of the skirt. With the center of the waistband at the natural waistline, the skirt length should not be more than one inch above or two inches below the crease in the back of the knee, and parallel to the ground.

The shirt should fit smoothly over the back of the shoulders, without wrinkles or tightness. There should be no gapping of the buttons across the bust, and the length of the shirt is determined by the individual's back waist length measurement. With few exceptions, the shirt will be correctly proportioned if it is properly fitted in other areas.

The shirt bottom should fit easily over the hips, conforming to the waistline and the sleeve of the shirt should extend to the bottom part of the wrist bone. The neck tab should be centered over the top button of the shirt, and the collar should lie smoothly on the neck without bulging and is not too tight.

Certain awards and accessories can be adjusted slightly on the uniform to conform to the individuals figure. When setting up the right side of the coat, center the bottom of the U.S. insignia disk on the right collar, approximately ⅝ inch up from the notch, with the center line of the insignia parallel to the inside edge of the lapel.

The unit crests should be centered on the shoulder loops, an equal distance from the outside shoulder seam and the outside edge of the button, with the base of the insignia pointed toward the outside shoulder seam. The regimental crest should be centered ½ inch above the nameplate and should be worn ½ inch above any unit awards or foreign badges, if worn.

The nameplate should be centered horizontally on the right side, between one and two inches above the top button. The placement of the nameplate can be adjusted to conform to the individual figure differences. The rank insignia should be centered between the shoulder seam and the elbow on both sleeves.

On the left side of the coat, the bottom of the branch insignia disk should be centered on the collar, approximately ⅝ inch above the notch, with the center line of the insignia parallel to the inside edge of the lapel. The shoulder sleeve insignia should be centered on the left sleeve, ½ inch below the top of the shoulder seam.

When special skill badges are worn above the ribbons, center them 1/4 inch above the ribbons. When more than one badge is worn above the ribbons, badges will be stacked ½ inch apart and may be aligned to the left to present a better appearance.

The ribbons should be centered on the left side with the bottom row parallel to the bottom edge of the nameplate. The marksmanship badges should be centered with the upper portion of the badge ¼ inch below the ribbons. If more than one marksmanship badge is worn, space them one inch apart. When special skill badges are worn below the ribbons, place them to the right of the marksmanship badges.

THE MEN'S CLASS A UNIFORM

The men's Army Green Uniform is authorized for year-round wear. The coat should fit snugly, without gapping, with ¼ to ½ inch of the shirt collar showing above the coat collar at the back, and fit smoothly over the back and shoulders. It should be slightly form-fitting around the waist (but not so snug that it causes the coat front to protrude) yet have some fullness and a draped effect. Four to six inches of looseness should be equally distributed around the circumference of the waist to provide comfort when moving.

The front of the coat should overlap without pulling or gapping. The buttons should present a straight vertical line. The coat should lay smoothly over the seat so that the back vents do not pull or gap. The bottom edge should be even and fall one inch below the bottom of the wrist bone and cover the shirt sleeves. The sleeve length should be one inch below the bottom of the wrist bone.

The trouser waistband bottom should rest on the hip bone plus or minus ½ inch. The waist should fit with a slight amount of looseness, but should not form a pleated effect when the belt is worn and tightened. The seat should fit loosely without wrinkling as the top of the leg blends into the seat. The bottom edge of the back of the trouser leg should fall midway between the heel top and shoe top. The bottom edge of the front of the trouser leg should rest on the middle of the instep and may have a slight break in the crease.

The tie should be tied so that it will fall in an area ranging from two inches above the top of the belt buckle to the bottom of the belt buckle. The tie can be worn in a windsor, half-windsor or four-in-hand knot and fit snugly and close to the neck.

The shirt collar should be smooth and lie close to the neck with about ½ inch of space between the shirt collar and neck. The shoulder seams should be at the mid-point of the shoulder, the back and shoulder should be smooth, without wrinkles or tightness. The front of the shirt should present a straight line of buttons, and there should be no evidence of gapping or pulling, and the "gigline" should be straight. The bottom of the sleeve should extend to the bottom part of the wrist bone.

The belt should have no more than two inches of extra webbing to permit adjustment.

When setting up the coat, the U.S. insignia should be placed approximately one inch above the notch center on the right collar, with the center line of the insignia parallel to the inside edge of the lapel. Place the bottom of the branch insignia disk approximately one inch above the notch, centered on the left collar, with the center line of the insignia parallel to the inside edge of the lapel.

The unit crests should be centered on the shoulder loops an equal distance from the outside shoulder seam and the outside edge of the button, with the base of the insignia pointed toward the outside shoulder seam. The regimental crest should be centered ⅛ inch above the top of the pocket flap. It should be placed ½ inch above the unit awards or foreign badges, if worn.

Unit awards should be worn ⅛ inch above the pocket flap, and the nameplate should be centered on the right pocket flap, between the top of the button and the top of the pocket.

The rank insignia should be centered between the shoulder seam and the elbow of the sleeves and the shoulder sleeve insignia should be centered on the left sleeve, ½ inch below the top of the shoulder seam.

When combat and special badges are worn, center them ¼ inch above the ribbons. When more than one are worn, stack them ½ inch apart and align them to the left to present a better appearance.

The ribbons should be centered ⅛ inch above the top of the pocket flap. One or more subsequent rows may be stacked ½ inch apart and may be aligned to the left to present a better appearance. The marksmanship badge should be centered on the pocket flap, ⅛ inch below the top pocket seam. If more than one badge is worn, they should be spaced one inch apart. When special skill badges are worn on the pocket flap, they should be placed to the right of the marksmanship badge.

The service stripes should be centered on the outside of the left sleeve, four inches from the bottom at a 45 degree angle. You can have one service stripe for each three years of completed service. The overseas bars are worn four inches up from the bottom end of the right sleeve and parallel to the bottom sleeve.

AWARDS IN ORDER OF PRECEDENCE

Medal of Honor
Distinguished Service Cross
Defense Service Medal
Distinguished Service Medal
Silver Star
Defense Superior Service Medal
Legion of Merit
Distinguished Flying Cross
Soldier's Medal
Bronze Star Medal
Purple Heart

Defense Meritorious Service Medal
Meritorious Service Medal
Air Medal
Joint Service Commendation Medal
Army Commendation Medal
Joint Service Medal
Army Achievement Medal
POW Medal
Good Conduct Medal
Army Reserve Components Achievement Medal
Army of Occupation Medal
National Defense Service Medal
Korean Service Medal
Antarctica Service Medal
Armed Forces Expeditionary Medal
Vietnam Service Medal
Southwest Asia Service Medal
Humanitarian Service Medal
Armed Forces Reserve Medal
NCO Professional Development Ribbon
Army Service Ribbon
Overseas Service Ribbon
Army Reserve Components Overseas Training Ribbon
United Nations Service Medal
Inter-American Defense Board Medal
Presidential Unit Citation
Joint Meritorious Unit Award
Valorous Unit Award
Meritorious Unit Commendation
Army Superior Unit Award
Philippine Republic Presidential Unit Citation
Republic of Korea Presidential Unit Citation
Vietnam Presidential Unit Citation
Republic of Vietnam Gallantry Cross Unit Citation
Republic of Vietnam Civil Actions Unit Citation

MORE FACTS ABOUT JUNIOR NCO PROMOTIONS

Corporals and Specialists can serve up to eight years prior to reaching their retention control point (RCP). They must complete the four week

Primary Leadership Development Course before they can be promoted. They must also meet the height and weight standard IAW AR 600-9 and pass the PT test before starting PLDC. A high school diploma or the equivalent of one (GED) is also needed.

A Sergeant can serve up to 13 years, but, if he or she is promotable, the time can be extended to 15 years. A Sergeant must attend a Basic NCO Course to qualify for promotion to Staff Sergeant. He or she must also meet the height and weight standard and pass the PT test before starting the course. The high school diploma or the equivalent is also required. NCOs attending the Basic course will have their length of service extended by one year.

6

SENIOR NCO PROMOTIONS (SFC, MSG, SGM)

CENTRALIZED PROMOTION

Headquarters, Department of the Army (HQDA) promotes soldiers to the rank of Sergeant First Class, Master Sergeant and Sergeant Major. Selection and promotion authority at HQDA does not deprive local commanders of the authorization to reduce soldiers in the rank of SFC, MSG or SGM for inefficiency or conviction by civil court.

The criteria for primary and secondary zones of consideration for each grade will be announced by HQDA before each board. A soldier may not decline the promotion consideration, but can decline to be promoted.

Selection by the DA board will be based on impartial consideration of all eligible soldiers in the announced zone. They are made by MOS and the best qualified in each MOS will be selected. The total number of soldiers selected in each MOS is the projected number the Army needs to maintain its authorized-by-grade strength at any given time.

Soldiers serving SRB or enlisted bonus (EB) service will not be promoted outside their CPMOS. If the soldier is in the zone of consideration for a HQDA promotion board, a "complete-the-record" evaluation report may be submitted according to the HQDA message announcing the zone.

Promotion certificates for soldiers promoted to SFC and above will be prepared by the unit or PAC for signature by the SGT or SSG promotion authority. Any higher level commander may direct that the signature authority be held at his or her level, but the certificates will still be prepared by the unit or PAC.

ELIGIBILITY CRITERIA

Before the DA board convenes, soldiers must meet the announced date of rank and basic active service date requirement and other eligibility criteria prescribed by HQDA.

They must have at least six (SFC), eight (MSG), and ten (SGM) years of total active federal service plus eight (MSG) and 10 (SGM) years of enlisted service creditable in computing basic pay for promotion to MSG and SGM. They must also be serving on active duty in an enlisted status on the convening date of the selection board.

Soldiers selected for promotion to SFC, MSG and SGM must have a high school diploma, a GED equivalent or an associate or higher degree. The soldier must not be ineligible to reenlist (IAW AR 601-280) for:

- AWOL during current enlistment without a waiver to reenlist
- Under approved local bar to reenlistment
- Qualitative Management Program
- Court-martial conviction during current enlistment without a waiver to reenlist
- Retirement
- Declination of continued service statement
- Flagged in accordance with AR 600-8-2

Soldiers must not have an approved retirement with a date prior to the convening date of the board or a signed DCSS. Staff Sergeants must be graduates of BNCOC or higher NCOES course in order to be considered for promotion to SFC. MSG that are removed from the promotion list for failing the Sergeant Major Course will not be considered by future promotion boards.

CENTRALIZED PROMOTION BOARD

The selection boards are composed of at least five members. They may be divided into two or more panels and each panel is composed of at least three voting members, including commissioned officers and senior NCOs.

The President of the board will be a general officer and an officer will be appointed to each board to serve as the recorder, but he or she will not vote. Female and minority groups will be represented.

The selection board will recommend a specified number of soldiers by MOS from the zone of consideration who are the best qualified to meet the needs of the Army. Soldiers who are not selected for promotion will not be provided specific reasons for not being selected. A separate memorandum of instruction will prescribe reports to be submitted, largest numbers to be selected, and other administrative details. The memorandum will be published as an enclosure to the memorandum announcing the results of the selection board.

PERSONAL APPEARANCE AND WRITTEN COMMUNICATION

No soldier may appear in person before a DA selection board on his or her behalf or in the interest of anyone being considered. The soldier's record is used to determine quality.

All eligible soldiers may write to the president of the promotion board to provide documents and information calling attention to any matter concerning themselves they feel is important to their consideration. Written communication is encouraged only when there is something that is not provided in the soldier's records which the soldier feels will have an impact on the board's deliberations. The proper address and due date will be included in the message announcing the zone of consider-ation. Correspondence will not be acknowledged, included in the OMPF or used as a basis for reconsideration.

Documents authorized for filing but are not in the soldier's OMPF should be sent to:

> Commander, USAREC
> ATTN: PCRE-FS
> Fort Benjamin Harrison, IN 46249-5301

Correspondence received from anyone other than the soldier concerned or that criticizes or reflects on the character, conduct or motives of any soldier, incomplete appeals for NCO-ERs, AERs, Court-Martials, Article 15s, etc., and copies of NCO-ERs will not be given to the board and therefore should not be sent to the enlisted record section. Non receipt of a letter to the board president does not constitute grounds for reconsideration or a standby advisory board.

BOARD RESULTS

HQDA will announce the results of a selection board by command memorandum. It will include a letter of instruction, board membership and the recommendation list. Names of soldiers recommended for promotion to SFC, MSG and SGM will be determined by seniority within recommended MOSs. Each MOS will have it's own selection list. Sequence numbers will be assigned within recommended MOSs based on seniority of date of rank, then basic active service date when DOR is the same, then age (oldest first) when DOR and BASD is the same.

The profile analysis provides insight into some of the areas that might have influenced the board's decision and the considered list is of soldiers considered for promotion by the board.

MONTHLY PROMOTIONS

HQDA will determine the total number of promotions to SFC, MSG and SGM on a monthly basis. The date of rank and effective date of promotion will be the same. PERSCOM will publish orders announcing promotion to SFC, MSG and SGM.

Prior to promotion, a SSG must be a graduate of ANCOC and MSGs must be graduates of the Sergeants Major Course. Conditional promotions are possible upon the successful completion of the required level NCOES. Soldiers who receive a conditional promotion will have their orders revoked and their names removed from the centralized promotion list if they fail to meet the NCOES requirement.

Any soldiers who fail to complete the required course for bona fide medical or compassionate reasons will not have their promotion revoked, but must pass the course to remain eligible for conditional promotions. Soldiers who failed the required NCOES course prior to 1 October 1993 will not be eligible for conditional promotion.

SERVICE OBLIGATION

Soldiers promoted to the grades of SFC, MSG or SGM will incur a 2 year service obligation. The obligation will be from the effective date of the promotion before non-disability retirement unless the soldier is eligible for retirement by completing 30 or more years active federal service, already eligible through prior service for a higher grade at the time of retirement, age 55 or older or expiration of their term of service.

Soldiers on a recommended list will be promoted on the last day of the month before being placed on the retired list if their sequence number has not been reached and they will complete 30 years of active service or have reached age 55.

Commanders will advise PERSCOM, (TAPC-MSP-E), Alexandria, Virginia 22332-0443 of these soldiers in time to allow for preparation of promotion orders before the soldier retires. A promoted soldier may not, at his or her own request, be reduced to terminate the required service obligation.

RECLASSIFICATION PRIOR TO PROMOTION

Soldiers on an SFC, MSG and SGM recommended list, who are reclassified prior to promotion, will receive a new sequence number within the new MOS based on their seniority, relative to other soldiers in the new MOS. Soldiers who have not been promoted will receive a letter through their PSC notifying them of the new sequence number.

The new number based on seniority will be assigned by taking into account all the soldiers selected for promotions in a particular MOS, whether promoted or not. If promotions have already occurred through the new number, the reclassified soldier will be promoted effective the first day of the month following the date of reclassification. If promotions have not occurred through the new number, the reclassified soldier will be promoted with contemporaries. If a soldier is promoted in an incorrect MOS and the promotion should not have occurred until a later date, the promotion will be revoked.

PREBOARD PROCESS FOR NCO'S IN ZONE OF CONSIDERATION

NCOs must meet the announced eligibility requirements for promotion board consideration and review and sign DA Form 2A and DA Form 2-1 for submission to the board. They should review their OMPF on microfiche four to six months prior to the board. Documents submitted for correction or additions to OMPF should be submitted through PSC to the U. S. Army Enlisted Records and Evaluation Center.

NCOs will be notified of procedures used to request consideration or reconsideration by the standby advisory board, if appropriate. Those that are not selected may be assisted by their BNS1 in writing to the career branch for promotion potential analysis. Documents supporting amendment, revocation, or late promotion orders must be received by the 20th of the month for action to be included in the promotion order booklet, that is mailed during the following month, to:

> Commander, PERSCOM
> ATTN: TAPC-MSP-E
> Alexandria, Virginia 22332-0443

STANDBY ADVISORY BOARD

The Deputy Chief of Staff for Personnel or designee may approve cases for referral to a standby advisory board, upon determining that a material error exists. The U. S. Total Army Personnel Command (TAPE-MSP-E) will determine if a material error existed in a soldier's OMPF when the file was received by a promotion board. An error is considered material when there is a reasonable chance that, had the error not existed, the soldier may have been selected.

Standby Advisory Boards are convened to consider records of those from the primary and secondary zones not reviewed by a regular board and those from a primary zone that were not properly constituted, due to a material error, when reviewed by the higher board. Standby Advisory Boards are also convened for recommended soldiers on whom derogatory information had developed that may warrant removal from a recommended list.

Soldiers selected by a SAB will be added to the appropriate recommended list and promoted alone with their peers when their seniority sequence number is reached. Only soldiers who were not selected from a primary zone of consideration will be reconsidered for promotion. A soldier that requests reconsideration will be given reconsideration for only one board.

Reconsideration normally will be granted when one or more of the following conditions existed on the performance microfiche (P-fiche) of a soldier's OMPF at the time it was received by a promotion selection board.

- An adverse NCO-ER or AER reviewed by a board was subsequently declared invalid in whole or part, and a determination was made that there was a material error.
- An adverse document belonging to another soldier is filed on the P-fiche of the non-selectee's OMPF and was seen by the board.
- An Article 15, administered on or after 1 September 1979, that was designated for the file in the MPRJ only, but was erroneously filed on the P-fiche of the OMPF and was reviewed by the board.
- An Article 15 punishment that was wholly set aside before 1 September 1979 was filed on the P-fiche when reviewed by the board.
- Court-martial orders were filed on the P-fiche of the OMPF when the finding was "not guilty".
- A document was filed on the P-fiche that erroneously identified the non-select as AWOL or a deserter.
- An MOS evaluation score or SQT score reviewed by a board and was subsequently recomputed by the evaluation center and resulted in a significant change.
- Receipt of a degree (AA, BA, BS) not recorded on the P-fiche or the qualification record, or was not seen in hard copy by the board. Only college degrees that are awarded by an accredited college or university (shown on official transcript) will be considered. The date of the degree will not be earlier than 3 months before the convening date of the board.
- An award of a meritorious service medal (initial award only) or higher, not recorded on the P-fiche or the qualification record or not reviewed in hard copy by the board. The date used for determination of reconsideration will be the date of the order or the effective date, whichever is later, and will not be more than 45 days before the convening date of the board.

- An annual or change of rater NCO-ER that was received at USAEREC early enough for processing and filing before the convening date of the promotion selection board was not received. NCO-ERs received at EREC prior to the PSC for administrative reasons may be a basis for reconsideration.
- An individual was considered in an MOS other than PMOS or CPMOS on the convening date of the board.

The ten items below do not constitute material error and will not be reasons for reconsideration:

1. Omission of letters of appreciation, commendation, congratulations or other similar commendatory correspondence.
2. Documents that are not derogatory having been filed on the wrong P-fiche.
3. Absence of documents written, prepared or computed following the convening date of the board.
4. Incorrect data on DA Form 2A or DA Form 2-1 reviewed by the soldier prior to the qualification record being forwarded to the board.
5. Absence of official photograph or the presence of an outdated one.
6. Absence of DA Form 2A and DA Form 2-1
7. Absence of an award lower than the Meritorious Service Medal.
8. Absence of documents not authorized to be filed in the P-fiche under AR 640-10 (i.e. ISRs)
9. Absence of completion of an NCOES course, unless it is a prerequisite for consideration and it was completed before the convening date of the board.
10. A complete the record NCO-ER under any circumstance will not be a basis for reconsideration.

All standby requests will be sent through the servicing PSC. Each case will be evaluated by the serving PSC. Cases clearly not meeting these guidelines will be returned. Correspondence, such as letters of commendation, appreciation and documents dated from third parties, will not be forwarded. Documents dated subsequent to the convening date of the requested standby advisory board will not be forwarded.

Requests for a standby advisory board will be sent to:

PERSCOM
ATTN: TAPC-MSP-E
Alexandria, Virginia 22332-0443

DA PHOTOS AND MILITARY RECORDS

As you know, you will not have to appear before the DA promotion board. They will go through your records, check out the last five of your NCO-ERs and look at your photo. Your record and NCO-ER will give them an idea of how you measure up to other NCOs in your grade and MOS. The only way they will be able to check out your appearance is by looking at your photo.

All NCO photos will be taken in color and they must be taken every five years. When I was still on active duty, I had one made every year, because I wanted the board to see the current me. On top of that, if they were looking at someone's photo that was three years old and compared it to mine that was just a year old (or less), I always had the feeling my photo would be the one that was more believable. A person doesn't change that much in a year, where as in five years, they may have gained weight or their uniform doesn't fit to standards. I would always sign my photo on the bottom of the back and put down my SSN. I reasoned that if records could get misfiled, why not a photo?

If a female soldier is due a photo during pregnancy, she will have six months after the birth of her baby to get one made and sent to DA.

When a DA photo is taken, regimental insignia will be worn alone with the branch insignia, all permanently authorized awards, decorations, combat and special skill badges and tabs. To take the best photo possible, it's better to take the picture in a heavy uniform because it photographs much better than a light weight one.

If that isn't possible, have the uniform cleaned and pressed just before you have your picture taken, and keep it in a long garment bag to keep the wrinkles out. Don't wear your uniform to the photographer, change it when you get there. Men should shave just before taking the photo and female soldiers should be aware that heels have a slimming effect when their picture is taken.

Most, if not all, Army installations now have contractors working the photo centers that take your photos. They will tape your uniform to help you look your best before taking the picture. Many NCOs and officers depend on this service, not knowing the regulation doesn't call for that

SENIOR NCO PROMOTIONS (SFC, MSG, SGM)

type of help.

The Army is now installing new technology where the soldier goes to the photo center, walks in the room to a predetermined spot and looks at a TV camera/monitor. What he or she sees is the same picture that will be seen by the Department of the Army. The only difference is that DA will get a hard copy of the photo, and the soldier will not. It will be up to the soldier to make sure the uniform is correct, but that's an NCO responsibility, anyway.

HQDA will send a message to the PSC announcing the zone of consideration. The PSC will send a letter to the BNS1 asking them to send the soldiers in the zone of consideration to the personnel office to review their records.

Be sure to take along the latest copy of your NCO-ER, physical and proof of any schooling or awards that may not be in your record. When you check out your DA Form 2A, pay close attention to:

- Your name and social security number
- Date of rank, PMOS and ASI
- PULHES and physical category code
- Date of last photo and NCO-ER verification
- PEBD/BASD and ETS
- Date of birth
- Ethnic and citizenship code
- Spouse social security number
- SMOS/ASI and duty MOS/ASI
- Date eligible for the next good conduct award.

Don't rush. Take your time and make sure everything is complete and accurate. When you check out your DA Form 2-1, pay close attention to:

- Name and social security number
- Overseas service (dates)
- Aptitude area scores (GT should be 110 or more)
- Awards, decorations and campaigns
- Civilian education and military schooling
- Appointments and reductions
- Specialized training (should also be on NCO-ER)
- Physical status (should be the same as on the PT score card)
- Personal and family data
- Current and previous assignments (update it)

When you check your records, don't forget the microfiche. This should be done every 6 months or whenever something is sent to DA to be placed in your records. To get a copy of your microfiche, send a signed, written request with your social security number and mailing address to:

> Commander
> U. S. Enlisted Records and Evaluation Center
> ATTN: PCRE-FF
> Fort Benjamin Harrison, Indiana 46249-5301

NCO-ER

In today's Army, there are many ways a soldier can be forced out of the Army. The number one way of being placed on the QMP list or forced out is the Non-Commissioned Officer Evaluation Report (NCO-ER). There are many soldiers out of the Army today, because of a bad report. Many receive the bad report because they knew no more about the NCO-ER than the rater or senior rater who cause them to be forced out or, in many cases, not get promoted.

In my book, *The Complete Guide to the NCO-ER*, I give detailed information about the NCO counseling checklist/record and the NCO-ER. There is information about counseling. Chapter three is all about bullet comments—why they are important, which are good ones and how to ensure you receive the best ones.

There is information about filling out the report and each block is covered, to include who fills out what and when. If a soldier should get a bad report, there is information about appealing the report. The book may be found at one of the post book stores or the clothing sales store. If not, look in back of this book under Military Titles to order.

MORE FACTS ABOUT SENIOR NCO'S PROMOTIONS

A Sergeant First Class can stay on active duty for a maximum of twenty two years and twenty four if he or she is promotable. A Master Sergeant can serve for a maximum of 24 years unless selected for promotion to Sergeant Major. A Sergeant Major can serve for a maximum of 30 years, but, can be selected to serve up to 35 years in a three or four star command position.

SENIOR NCO PROMOTIONS (SFC, MSG, SGM)

The average length of service, as of 1994, is thirteen for promotion to Sergeant First Class, seventeen to Master Sergeant and twenty years to Sergeant Major. If a soldier fails the Sergeant's Major Academy, he or she can't return to the course or compete for promotion, again. If he or she were promoted to Sergeant Major prior to going to the course but fail, they will be reduced to Master Sergeant and will not be able to compete again, although they will incur a two year obligation for attendance at the course.

Promotable MSG's can apply to be flocked to Sergeant Major one day before assuming the position if he or she:

- Is selected for promotion to SGM
- Is a graduate of the SGM course
- Will serve in a SGM position
- Need the rank to perform the duties of a SGM

NCOs can't just sit around with their fingers crossed hoping they will get promoted. Most likely they will get QMP'd first. NCOs need to:

- Get their reading level up to the 12th grade
- Get an Associate Degree as soon as possible
- Take the hardship assignments
- Get a DA photo each year (although the Army requires one only every five)
- Always get a good NCO-ER
- Get and keep key leadership positions
- Learn how to operate a computer
- Keep fit, pass the PT test and meet height/weight standards
- Plan and manage their careers

7

REMOVAL AND REDUCTIONS

REMOVING A SOLDIER FROM THE STANDBY PROMOTION LIST

A soldier will be given written notification at least 15 working days prior to the date of the board. The board will be composed of an unbiased membership and the recorder will arrange for any reasonable available witnesses the soldier wishes to call on his or her behalf. Copies of all written affidavits and depositions of witnesses who are unable to appear before the board will be furnished by the soldier. Army Regulation 15-6 does not apply to removal boards.

The soldier has the right to decline, in writing, to appear before the board or appear in person during all open proceedings for cause. He or she may challenge any member of the board and request any reasonable available witnesses whose testimony he or she believes to be pertinent to the case.

The soldier will state in his or her request the type of information the witness will provide. He or she may present written affidavits and depositions of witnesses who are unable to appear and may elect to remain silent, to make an unsworn or sworn statement, or be verbally examined by the board. The soldier may also question any witness appearing before the board.

Failure of the soldier to exercise his or her rights will not negate the board's proceeding, finding or recommendation.

The President of the board will ensure that enough testimony is presented to enable the board members to fully and impartially evaluate each case and arrive at a recommendation. He or she will prepare a report, in writing, of the board proceedings and submit it to the promotion authority.

The promotion authority may direct a new board if an error in the conduct of the board had a material adverse effect on an individual's substantial rights. If a board failed to consider all available evidence in the case, a new board may also be directed.

REMOVAL FROM A CENTRALIZED PROMOTION LIST

To have a soldiers name removed from a DA centralized promotion list, the commander must request removal of the soldier by requesting the BNS1 to prepare DA Form 268.

The BNS1 will then verify that conditions for removal have been done according to the regulation. They will prepare, authenticate and forward the DA Form 268 to the PSC and a copy to Commander, PERSCOM, ATTN: TAPC-MSP-E, Alexandria, Virginia 22332-0443.

Removal documentation will be prepared and forwarded to the commander for his or her signature. The commander will sign the request and inform the NCO, in writing, of the intent to remove him or her from the promotion list and return the action to the BNS1. The BNS1 will then review the action, obtain the Battalion Commander's signature and forward it, through command channels, to the PSC promotion section.

Once the PSC's promotion section receives the action, they will review it for compliance, obtain the concurrence or non-concurrence of the General Court Martial Convening Authority (GCMCA) or the first general officer in the chain of command that has a Staff Judge Advocate. Once the GCMCA has approved the action, the PSC will inform, in writing, the NCO of the action.

The soldier then will be given fifteen working days to respond to the action. If the soldier decides not to respond, he or she will sign a statement to that effect.

The action will then be forwarded, in duplicate, to PERSCOM, to include a copy of the DA Form 268 and the soldier's DA Form 2A and 2-1. When the BNS1 receives the response form PERSCOM, they will forward it to the command and ensure a copy is filed in the MPRJ and prepare final DA Form 268s and forward a copy to PERSCOM. The Commander will notify the soldier of the outcome.

PROCESSING COMMAND INITIATED REMOVALS

Commanders may recommend that a soldier's name be removed from a HQDA recommended list at any time, but the recommendation must be documented fully.

When recommending a soldier for removal punishment under the UCMJ or nonpunitive measures, it should not automatically be the sole basis to suggest that a soldier's name be removed from the recommended list. The soldier's conduct, before and after the punishment or nonpunitive measures and facts and circumstances leading to and surrounding the misconduct must be considered. To remove a soldier, based solely on a minor or isolated incident of misconduct, may be unfair to the soldier.

Removal from a HQDA list has far-reaching, long-lasting effects on the soldier. The probability for subsequent selection for promotion is extremely limited. Commanders should evaluate circumstances to ensure that all other appropriate actions have been taken (training, supervision, formal counseling have not helped) or the basis for considering removal is serious enough to warrant denying the individual's promotion.

The Commander must submit a recommendation for removal on a soldier who is not in compliance with the 6 or 12 month rule in AR 600-9. Recommendation may be submitted for substandard performance, but, should not be submitted on isolated acts, based on short-term supervision. Each case must be investigated thoroughly.

The removal action will be submitted for review, through command channels, to the Commander having General Court-Martial convening authority or the first General Officer in the chain of command having a Staff Judge Advocate on his or her staff. The GCMCA or an Army General Officer will make recommendations and sign the removal action. Before sending a removal action to HQDA for consideration, it will be

REMOVAL AND REDUCTION

delivered, in writing, to the soldier concerned with all documents that are to be forwarded to HQDA included with the notification.

The soldier will be allowed to respond to the proposed action, and may submit a rebuttal within fifteen working days after receipt of the written notice. A soldier who elects not to respond will sign a statement that he or she has reviewed the proposed action and elects not to respond.

After that soldier has submitted a rebuttal or signed a statement to the effect that he or she does not elect to submit a rebuttal, the action will be sent to PERSCOM in duplicate and will include a copy of DA Form 2A, DA Form 2-1 and DA Form 268. HQDA will make the final decision on the removal, based on results and recommendations of the DA standby advisory board.

HQDA INITIATED REMOVALS FROM A CENTRALIZED PROMOTION LIST

HQDA will continuously review promotions lists against all information available to ensure that no soldier is promoted where there is cause to believe that a soldier is mentally, physically, morally or professionally unqualified to perform the duties of the higher grade.

A soldier may be referred to a standby advisory board (STAB) for:

- Punishment under Article 15, UCMJ (whether filed in the restricted fiche or performance fiche of the OPMF)
- Any court-martial conviction or a memorandum of reprimand signed by a General Officer and placed in the OMPF
- Adverse documentation filed in the OMPF
- Failure to meet the requirements for weight standards as outlined in AR 600-9
- Other derogatory information received by HQDA, but not filed in the OMPF

Suspension of favorable personnel actions, failure to initiate a suspension of favorable personnel actions, however, does not invalidate referral of the action to the STAB or subsequent actions relating to the recommendation of removal. A soldier referred to a STAB normally will be considered by the STAB within 120 days after the case is referred for review.

A soldier who has been provided with 30 days from the date of receipt of such information to submit comments is considered to have been provided a reasonable opportunity, unless a good cause is shown.

The STAB will consider the official military personnel record, consisting of the OMPF (including relevant portions of the restricted fiche), enlisted records brief and an official photo. Additional information by HQDA, but not on file in the OMPF, which the authority finds is substantiated, relevant and might reasonably and materially affect a promotion recommendation, provided such information has been properly referred to the soldier for comment. The soldier may include the opinion and statements of third persons with his or her submission.

A soldier who is removed from a promotion list may appeal that action only in limited instances. Action will be taken only by HQDA on appeals. Soldiers may appeal an action to remove them from a centralized promotion list when the underlying action(s) that formed the basis of the removal are subsequently determined to be erroneous, based on facts that were not available or reasonably discoverable at the time of the original action (at the time the soldier was notified of the removal action) or for other compelling reasons.

Appeals should be referred through command channels, to include GCMCA, to HQDA (TAPC-MSP-E). The action will be no later than fifteen days after the commander is notified of the appeal.

REDUCTION IN GRADE

A reduction board is required for soldiers in the grade of SGT through SGM for any reduction for misconduct and inefficiency. Board appearance may be declined in writing, which will be considered as acceptance of the reduction board's action.

A court-martial sentence of soldiers which, as approved by the convening authority, includes a punitive discharge, confinement or hard labor without confinement, carries a reduction to the lowest enlisted pay grade. The convening authority must suspend execution of sentence and provide, in his or her action, that the soldier will serve in that grade during the period of suspension and thereafter, unless the suspension is vacated before its termination.

When the separation authority determines that a soldier is to be discharged under other than honorable conditions, they will be reduced to the lowest enlisted grade.

Reduction transactions, for all grades, are submitted by the PAC. The legal clerk at Battalion level will maintain an Article 15 log. A control copy of the Article 15 will also be maintained at the Battalion level and verified by the PAC supervisor when a reduction transaction is required.

REMOVAL AND REDUCTION

Only Article 15s containing forfeitures will be forwarded to the finance center.

REDUCTION FOR MISCONDUCT

A soldier convicted by a civil court or adjudged a juvenile offender by a civil court, will be reduced or considered for reduction on receipt of documents establishing a sentence or a finding of guilty with sentence to be established at a later date. Juvenile offender includes adjudication as a juvenile delinquent, wayward minor or youthful offender. A soldier may be reduced even though an appeal is pending or has been filed.

If a reduction board is required, the reduction action must be accomplished within thirty duty days after receipt of documentary evidence and before separation or retention is considered. The reduction authority may extend the thirty day limitation for good cause. A written justification must be included in the file if an extension is granted.

Commanders will publish orders and enter the reduction in the military records of the soldier. The authority for the reduction will be cited in the order. The soldier will be notified in writing of the right to appeal the reduction. The written notification will include the time limits and procedures for an appeal.

If the conviction is reversed, the soldier will be restored to the former grade. If the sentence is modified or reassessed, action will be taken and the soldier will be notified in writing of the decision. If a soldier is reduced prior to sentencing and the sentence imposed is less severe, the soldier will be notified, in writing, of the decision.

Soldiers in the grade of SFC through CSM cannot be reduced for an Article 15 of the UCMJ under this provision.

REDUCTION FOR INEFFICIENCY

A demonstration of characteristics that show that a person cannot perform duties and responsibilities of the grade and MOS is inefficiency. It may also include any act or conduct that clearly shows that the soldier lacks those abilities and qualities normally required and expected of an individual of that grade or experience.

Misconduct, including conviction by civil court, as bearing on efficiency may also be considered by commanders. A soldier may be reduced under this authority for long standing unpaid personal debts that he or she has not made a reasonable attempt to pay. A soldier must have served in the same unit for at least 90 days prior to being reduced one grade for inefficiency.

The commander starting the reduction action will present documents showing the soldier's inefficiency to the reduction authority, which may include, statements of counseling and documented attempts at rehabilitation by the chain of command or supervisors, record of misconduct during the period concerned, correspondence from creditors attempting to collect a debt from the soldier and adverse correspondence from civil authorities. Documents should establish a pattern of inefficiency rather than identify a specific incident.

Reduction for inefficiency will not be used to reduce soldiers for actions in which they have been acquitted, because of court-martial proceedings or in place of an Article 15 of the UCMJ. To reduce a soldier for a single act of misconduct, the Commander reducing the soldier will inform him or her, in writing, of the action contemplated and the reason. The soldier will acknowledge receipt of the memorandum by endorsement and may submit any pertinent matter in rebuttal.

CONDUCTING A REDUCTION BOARD

A soldier who is to appear before the board will be given at least 15 duty days written notice before the date of the hearing. The soldier or his or her counsel must have time to prepare the case.

If the soldier requests counsel, the convening authority will determine if a judge advocate or military counsel is reasonably available. If a judge advocate is available, the request will be forwarded to the local trial defense service official for necessary action. The availability of the judge advocates will be determined by the requested individual's trial defense service supervisory official. Determinations as to the availability of the judge advocates or named counsel are final.

Notification of a board hearing date will be made after counsel is available, the recorder, on request, will arrange for the presence of any reasonable available witness or witnesses the soldier desires to call on his or her behalf. Copies of all written affidavits and depositions of witnesses who are unable to appear before the board will be furnished the individual or his or her counsel as appropriate.

The President of the board will ensure that enough testimony is presented to enable the board member to fully and impartially evaluate each case, be objective in their deliberations, arrive at a proper recommendation, consider those abilities and qualities required and expected of a solder of that grade and experience and determine the best interest

REMOVAL AND REDUCTION

of the Army. Army Regulation 15-6 does not apply, except when determining the availability of judge advocates.

The board may recommend reduction of one or more grades, reassignment in grade or retention of current grade. A retention in the current grade recommendation may include a recommendation that the soldier be removed from a HQDA or local recommended list. The board members cannot recommend a lateral appointment. A majority of the appointed members of the board will constitute a voting quorum and must be present at all sessions.

The convening authority may approve or disapprove the severity of the board's recommendation. Unless suspended by the convening authority, approved reduction recommendations are effective immediately, without regard for appeal procedures.

If a soldier is being reduced for inefficiency, the convening authority may direct suspension of the reduction for a period not to exceed 6 months. If the suspension is not vacated during that period, reduction can only be accomplished by convening a new reduction board.

A recommendation to remove a soldier's name from a HQDA recommended list will be forwarded by the convening authority to the GCM convening authority or the first Army General Officer commander who has a staff judge advocate or legal advisor available, who will review the proceedings and take final action. PERSCOM will administratively remove the soldier's name from the recommended list. If the approving authority does not agree with the recommendation, the action will be returned through command channels to the convening authority with the reason for the disapproval.

If a civil conviction is reversed, the soldier will be restored to the grade from which reduced.

SOLDIER'S RIGHT TO APPEAL

Failure of the soldier to exercise the right to counsel will not negate the board's proceeding, finding and recommendation.

A soldier may decline, in writing, to appear before the board or may appear in person with or without counsel at all open proceedings. He or she should respond, in writing, within seven duty days of notice by the reduction authority stating his or her desire to appear, or not appear, before a reduction board.

A soldier may retain a civilian lawyer at no expense to the government. If not represented by a civilian lawyer, the soldier may request the appointment of a detailed judge advocate, an appointment of a named military counsel, or a detailed military counsel.

The soldier will be advised by the board president of the nature of the action being contemplated, the impact of such action on continued military service, and the right to request counsel. The soldier may challenge any board member for cause, or request any reasonably available witness whose testimony is believed to be helpful to the case. He or she will explain the nature of the information the requested witness should provide.

The soldier may submit written affidavits and depositions of witnesses who are unable to appear before the board. The soldier may also employ provisions of Article 31 of the UCMJ or submit to an examination by the board. The soldier or counsel may question any witness appearing before the board.

Soldiers submitting appeals will be informed, in writing, of the decision. A copy of the appeal and the final action will be provided to the custodian of the soldier's MPRJ. Authority to take final action on an appeal may not be delegated.

8

NCO EDUCATION SYSTEM (NCOES)

REASON FOR THE NCOES

The NCO education system is a series of schools the NCO must attend before getting promoted to SGT, SSG, SFC, MSG or SGM. The system is used to train and develop NCOs as they progress up the chain to the rank of SGM.

The NCOES is a must for those soldiers that want to get promoted and/or remain on active duty. The courses are:

- **Primary Leadership Development Course** (PLDC) for promotion to Sergeant

- **Basic NCO Course** (BNCOC) for promotion to Staff Sergeant. The course consists of two weeks of leadership material and several weeks of MOS specific material that varies in length depending on MOS.

- **Advanced NCO Course** (ANCOC) for promotion to Sergeant First Class. It's similar to the basic course, but the material is tailored for NCOs preparing for higher levels of responsibility.

- **Sergeant's Major Course** (SGMC) for promotion to Sergeant Major. The course prepares NCOs for the highest level or responsibility in tactical units and headquarters. It's a nine month (PCS) course held at Fort Bliss, Texas.

FITNESS BEFORE TRAINING

Soldiers are weighed and, if necessary, taped for body fat estimate as soon as they arrive for one of the NCOES courses, then they must pass the Army Physical Fitness Test. If they don't meet the weight standard or pass the PT test, they will have to return to their unit.

Physical fitness in an NCOES course is to teach the soldier how to instruct and keep their soldiers physically fit, not to get themselves fit during the course. One of the reasons some of the soldiers are returned to their unit is because they are not doing the push-ups and sit-ups by the book at the unit, and are not properly prepared when they report for one of the NCOES courses, and fail the test.

The two mile run is another reason some fail the test. Many units run in a unit formation, where all the soldiers run at the same pace. If they all run the two miles in 18 minutes but need to run it in 16 minutes to pass the test, then the unit run is not doing them any good. The same thing holds true for the timing of the push-ups and sit-ups.

Even though these may be reasons soldiers fail the physical fitness test, these are not reasons to be admitted to the NCOES course should a soldier fail the test, because each soldier is responsible for his or her weight control and physical fitness.

READING BEFORE LEADING

Reading ability is very important to the soldier and is one of the main reasons they should be tested for their reading and comprehension level before they go to an NCOES course.

The reading level is measured by the Tests of Adult Basic Education (TABE). Soldiers who don't meet the reading level for their rank should be given remedial courses and tested again, until they score at the required level. Soldiers should score at the 10th grade level in vocabulary and comprehension before attending the primary, basic and advanced courses, and at the 12th grade level before attending the Sergeant Major Course. A speed-reading course would also be helpful before attending the Sergeant Major Course.

REQUIREMENT FOR PLDC ATTENDANCE

- Must be a SPC (P) for primary or SPC for alternate
- Must have no temporary profiles
- Must meet height and weight standards IAW AR 600-9
- Must pass the Army Physical Fitness Test
- Must be recommended by the unit commander
- Must have 180 days of service remaining after the course
- Must not be barred from reenlisting
- Must have serviceable clothing and equipment
- Must bring his or her protective mask, complete with accessories
- Must bring meal card or DD Form 1610 with per-diem marked "No"

COURSE INSTRUCTION FOR PLDC

- Leadership
- Counseling
- Military customs and courtesies
- Wearing of the uniform
- Drill and ceremonies
- Physical readiness training
- How to plan and conduct inspections
- Effective communications
- Family team building
- Equal opportunity/Sexual harassment
- Vehicle maintenance; PMCS supervision
- Supply economy
- Team development
- History of the NCO
- Train the force (how to prepare and give a class)
- Land navigation/map reading
- Environmental awareness
- Field sanitation

- Basic rifle marksmanship
- Combat orders
- NBC/96 hours STX
- MILES for small arms weapons

REQUIREMENT FOR BNCOC ATTENDANCE

- The soldier must be recommended by the commander
- Must have passed the APFT in the past six months
- Must meet physical fitness and height/weight standards
- Must have 180 days remaining after graduation (letter of intent does not qualify)
- Must be on unit order of merit list
- Must have received a six week notice about schooling
- Must have been counseled about the importance of the course/career
- Must have military/civilian glasses if authorized
- Must have meal card or DD Form 1610 (no per-diem)
- Permanent profile soldiers must have MMRB results to enroll
- Commandant must be notified of temporary profile soldiers
- Must have TABE test results in their possession
- May have sponsor to assist during the course
- Must have necessary clothing and equipment for the course
- Must be on the promotion standing list (have a copy to enroll)
- Class "A" uniform must fit correctly and insignia properly placed
- Must have copy of DA Form 2A and 2-1
- Must have protective mask, complete with accessories
- Must have passed pre-math test (some MOSs)
- Must be a PLDC graduate and served in the unit six months between courses.

COURSE INSTRUCTION FOR BNCOC

- Army writing
- Military leadership
- NCO-ER
- Risk management
- NBC defense

- Combat orders
- Troop leading procedures
- Squad tactical operations
- Battle focus training
- Rifle marksmanship
- PFT's
- Equal opportunity
- Army team building
- Army family building
- Property accountability
- PERS/PERF COUN.
- Diagnostic MOS overview
- BRM training
- LADP
- Training the force
- CLT Exam

Some of the MOS type courses will change, depending on the MOS of the soldier taking the course. The NCO must also pass the PT test before being able to enroll in the BNCOC.

REQUIREMENT FOR ANCOC ATTENDANCE

The Advanced NCO Course is very similar to the basic course in that there is an administration part and a MOS portion that promotable Staff Sergeants must complete. They must also meet the height and weight standards and pass the PT test before starting the course.

It would be helpful to the NCO to be able to read at a 10th grade level before reporting for the course. The subjects covered at the advanced course will depend on the NCO's MOS.

COURSE INSTRUCTION FOR ANCOC

- Effective writing
- Military leadership
- Development leadership assessment
- NCOER/Counseling
- Combat orders
- Troop leading procedures

- Platoon tactical operations
- Battle drills
- Risk management
- Marksmanship training
- Physical fitness training
- Equal opportunity
- Army family team building
- Military justice
- Law and warfare
- Battle focused training
- Environmental awareness

REQUIREMENTS FOR FSC ATTENDANCE (NOT PART OF NCOES)

The First Sergeant Course (FSC) is a twenty five day, fast paced, interesting and challenging course.

The type of instruction used at the FSC is the small group process. The class will be divided into fourteen to sixteen member groups which will have a mixture of MOSs and backgrounds. There will be nine or ten groups for the entire five weeks.

There is no platform type instruction, learning takes place through an exchange of information and ideas among group members. The lessons are fast paced. The instructor guides the discussion in the classroom, but the students learn from the books and each other's experience.

There is a study hall for each group. All tests are open-book and the answers are usually word-for-word from the manuals. The NCO has the additional responsibility to develop and conduct physical fitness training.

The major lesson modules taught in the course are:

- Unit personnel management and administration
- Leadership, discipline and morale
- Logistics, maintenance and security
- Physical readiness, operations and training

The uniform is normally BDUs. The Class "A" uniform will be used for the class photos and graduation. Light weight civilian clothing is recommended for summer weather, because El Paso is generally hot and dry. It is cool, dry and windy in the winter.

Students are not authorized to bring their family, unless so stated on their orders. Adequate quarters are available, therefore a certificate of non-availability will not be issued. The quarters rate is about $16.00 per night, the billeting office is open 24 hours and will accept cash, personal checks, MasterCard, Visa and the Army Diner's Club card.

Students should draw a maximum per diem payment before leaving for the FSC, because Fort Bliss FAO will only pay 80% of your authorized per diem rate. All Sergeants First Class and above can participate in the Diners Club credit card program, or they can pay up front and settle their claims upon returning to their duty station.

Do not take any type of privately owned weapons to the FSC, unless stated otherwise on your PCS orders. Keep your health and shot records in your possession at all times; this is one of the reasons you will need a brief case. Your instructor will issue the required materials to you on the first day, therefore, take a box or laundry bag with you.

Ordinary leave will not be granted while in a student status. If you drive your POV, you must temporarily register it at Bldg. 117 (on Main Post), so have your proof of insurance. If you ride a motorcycle, you should contact the Fort Bliss safety officer for proper safety gear requirements at DSN 978-5611/2510, commercial (915) 568-5611/2510.

There is limited transportation from Biggs Field (FSC location) to and from Main Post, other than the buddy system, but the FSC does provide a 15 passenger van for your use to the main post locations, such as the PX, Commissary, clothing sales store, etc. If you have a DDC card that was issued within the last four years, take it with you when you in-process.

Protestant services are available on Biggs Airfield, with other denominational services available on Main Post.

Here is an approximation of how much money you may need during the course:

Daily quarters cost	=	$16.00
Daily Meals/Incidentals allowance	=	$18.15
Daily Total	=	$34.15
TDY Days		x 35
TOTAL	=	**$1195.25**

Voluntary incidental costs at the course are:

- Class Ring = about $110 plus
- Class Photo = about $15.00
- Class Coin = about $5.00

- Plaque Mounted Diploma = about $40.00
- NCO Lifetime Museum Membership = $25.00

Your mailing address will be:

Your Name
FSC and Course Number
U. S. Army Sergeant Major Academy
Fort Bliss, Texas 79918-1270

You can reach the senior instructor and company commander at DSN 978-8205/8419 or (915) 568-8205/8479. After duty hours call DSN 978-8081 or (915) 568-8081. Just like courses in the NCOES, the students must satisfactorily meet the following standards:

- Academic standards
- Physical fitness standards
- Body fat standards
- Conduct and discipline standards

All students will receive an academic evaluation report (AER), regarding their accomplishments, potential and limitations during the course. Before attending the First Sergeant Course:

- You must be medically cleared to take the APFT
- You must take the APFT unless prohibited by a medical profile if you reached the age of 40 on or after 1 January 1989
- You will not be enrolled if you have a temporary profile for all three events
- You will not be enrolled if you fail the PT test or height/weight standards
- You will need 20 copies of your TDY orders and two copies of receipt for advance pay
- You will need a DA Form 30 if you were on leave before the course or are going on leave after the course.
- You will need your MPRJ, if you're attending the course TDY en route to a PCS.
- If age 40 or older you must bring a copy of latest proof of clearance by DVST. (DA Form 4970 or DA Form 4070E)
- NCOs with permanent physical profiles must have DA Form 3349 (Medical Condition)

NCO EDUCATION SYSTEM (NCOES)

- You must have a PT score card from your unit showing your last score.
- You must have a copy of your DA Form 2, 2-1, 93, and VA Form 29-8286 which is the SGLI insurance
- You must have one Class "B" uniform and three sets of BDUs
- You must have a complete Class "A" uniform, low quarters and socks.
- You must have Field Jacket and gloves w/inserts if you attend October-April
- You must have a PT uniform (grey shirt, shorts) and white calf-length socks. Black gloves, watch cap and sweats are needed for winter months.
- A windbreaker and/or pullover sweater is optional.

Below is a list of recommended reading prior to attendance at the FSC:

FM 22-100	FM 22-103	FM 101-5
AR 600-200	DA Pam 600-8	DA Circular 623-88-1
AR 614-200	AR 600-8-2	AR 601-280
AR 672-5-1	AR 37-104-3	AR 25-55
AR 340-21	AR 220-45	AR 630-10
AR 700-84	AR 635-200	FM 22-600-20
AR 600-20	FM 22-100	DA Pam 608-41
AR 40-3	AR 614-30	DA Pam 608-47
AR 630-5	AR 600-85	DA Pam 600-24
FM 22-101	FM 27-14	FM 27-10
AR 670-1	FM 22-5	AR 612-11
AR 750-1	AR 43-5	DA Pam 738-750
AR 710-2	AR 735-5	DA Pam 612-1
TB 710-5	AR 600-38	FM 10-23
AR 190-11	AR 190-31	AR 190-5
DA Pam 190-51	FM 21-20	DA Pam 350-18

Bring a copy of FM 25-101 to the FSC with you along with a copy of your unit or battalion METL and unit mission statement. Don't let the recommended reading list turn you off, there is no reason to read the complete manuals, unless you just want to, because most of the subjects covered at the FSC are only one to three hours long.

COURSE INSTRUCTION FOR THE FSC

- Introduction to soldier team development
- Introduction to communication skills
- Troop leading
- Role of the First Sergeant
- Professional ethics
- Stress management
- Army family action plan
- Alcohol and drug prevention and control program (ADAPCP)
- Equal opportunity program
- Suicide prevention
- Counseling (personal/performance)
- Manual for courts-martial
- Soldier's rights
- Pre-trial confinement
- Article 15 (UCMJ)
- Military appearance
- Military customs and courtesies
- Unit sponsorship program
- Developmental leadership assessment
- Airland battle doctrine
- Combat orders
- Unit training
- Unit displacement operations
- Tactical sustainment operations
- Tactical recovery operations
- Field preventive medicine
- Casualty evacuation
- Individual physical training
- Unit physical fitness program
- Weight control program
- Drill and ceremonies
- Unit manning report
- NCO evaluation reporting system
- Personnel and Finance actions
- Enlisted promotions and reductions
- Unit retention program
- Awards and decorations

- The PAC system/ Personnel Service Support
- Unit standing operating procedures (SOP)
- Privacy act
- Duty rosters
- Unit preventive maintenance program
- Military property accountability
- Supply management
- The Army Field Feeding System (AFFS)
- Physical security
- AWOL/DFR actions
- Battle Focused Training

REQUIREMENTS FOR SGM ACADEMY ATTENDANCE

The new nine month Sergeant Major Academy is the best in the Army.

Like the other leadership courses, the NCOs must meet the height and weight standards and pass the physical fitness test to start the course. It would also be helpful to them if they could read at a 12th grade level and know how to operate the Army computers. A speed reading course will also be helpful to the NCO.

During the course, all students will earn an Associates Degree or higher, depending on how much college education they have before the course. Along with the PT test the students will have to pass to begin the course, they will have to take another one six months after starting the course, and those NCO who choose to, may take the Master Fitness Course and earn the ASI for it.

There was a time that if a Sergeant Major was later selected to be a Command Sergeant Major, he or she had to return to Fort Bliss for a three week course to become one. Now, with the new nine month course, all students will take the Command Sergeant Major part and receive the ASI for it, even if they are promoted to Staff Sergeant Major. Battle tactics are the main reason the course was extended to nine months, and will give the NCO another ASI. There will also be a speed reading course, as well as a computer operator course for those that need it.

As the students go through the course, their spouse will also learn and see what they are doing, so that they can better understand why the Sergeant Major position is so rewarding and demanding.

Of all the benefits the students will receive, the best one is that they will all be promoted to Staff Sergeant Major or Command Sergeant

Major after the course.

After completing the course, the students must serve an additional 18 months, but this could be extended to two years, because of all the benefits. If you think about it, there are not many places where you can get an Associates Degree in nine months, free of charge.

COURSE INSTRUCTION FOR THE SGM ACADEMY

All NCOs that attend the Sergeants Major course will be trained as Division Command Sergeants Major. As they sign up for the course, each will be assigned to a make believe division, and it will be their job to form, train, activate, deploy and deactivate the division. To form or get the division started, the students will have to:

- Request soldiers for the complete division
- Order all TO&E equipment
- Order supplies, manuals and training aids
- Equip all levels from division to the smallest unit

Once the students are told their division is completely built, their next step will be to train the division, to include:

- Start training for the mission
- Battle focused training
- Develop mission essential tasks lists
- Maintenance programs
- Training planning
- Mobility, counter mobility and survivability
- Long, short and near-term planning
- Command and control
- Combat service support
- Air defense
- Fire support
- Maneuver
- Safety
- Environmental law
- Soldier care
- Family support groups
- Taking care of the Army family

- The elements of the Army family team building program
- Army philosophy concerning the total Army family
- Factors influencing the welfare and well-being of soldiers and their families
- Army families benefits and entitlements
- Leader responsibility
- Function and purpose of the family support group
- Function of a rear detachment as it pertains to a soldier's family
- The Army culture
- DOD policy on homosexual conduct
- Sexual harassment/Equal opportunity

After all the units in the division are fully trained, the students will be told that the division is to be activated and placed on alert, in preparation to deploy. The students will be told what country or region they may have to deploy and they will have to do a staff study about that region, to include:

- The United States' relationship with the region
- How they feel about Americans
- The weather and type of terrain
- Their military strong and weak points
- Soldiers and equipment needed for the deployment
- The best time to deploy
- The time it takes and the best way to get there
- How long the mission should last
- Equipment needed for advance teams
- Back up plans of attack

After the staff studies are complete and graded, the students will be told to deploy their division and prepare to fight, if need be. The student will plan the deployment, to include:

- Arranging transportation for soldiers and equipment
- Integrating the National Guard with division soldiers
- Getting equipment to the rail-head, ships and airports
- Identifying pick-up point
- Deploying the division
- Setting-up new location in the region
- Accounting for all soldiers and equipment
- Starting command and control with junior units

- Replacing equipment and soldiers as needed
- Keeping information updated on the enemy
- Emplacing weapon systems and fighting teams as needed
- Setting-up POW camps as needed

When the division deploys to the region and has accomplished the mission, the students will be told to return the division and deactivate it, to include:

- Return the National Guard to their home units
- Return the division soldiers and equipment
- Service and turn in all equipment
- Retire and PCS soldiers as needed
- Retire the Division Colors and flag
- Conduct awards and retirement ceremony

There you have a small list of what the course will be about. Another way to put it is, the student will have to do everything it takes to form a division, train the force, go to war and return and deactivate it.

Most Sergeants Major start at the Battalion level, but there is nothing that says the Division Commander can not start them at the Division level because of their training.

APPENDIX A

NON PROMOTABLE STATUS

If any NCO (SFC-SGM) whose name appears on a HQDA recommended list are non promotable, the Commander is responsible for notifying HQDA.

If AR 600-8-2 applies, the Commander must forward documentation, to include the initial DA Form 268 (Report to Suspend Favorable Personnel Action), and the reason for the flagging. If the flagging action is closed, the PSC will forward a copy of the final DA Form 268 to HQDA, the date the flag was closed, the punishment received, the date all punishment is completed and/or the date that a letter of reprimand was actually imposed.

For all other cases, the PSC will provide HQDA with the soldier's name and a brief summary of circumstances that caused the soldier to become nonpromotable. The PSC will send a copy of the counseling statement for solders (SSG-MSG) holding promotion recommended list status to HQDA. All correspondence should be sent to:

Commander, PERSCOM
ATTN: TAPC-MSP-E
200 Stovall Street
Alexandria, Virginia 22332-0443

This correspondence will include the NCOs sequence number, promotion military occupational specialty (PRMOS) and date the NCO became non-promotable. Soldiers (PV1-MSG) are non-promotable to a higher grade if one of the following conditions exist:

- A local security violation

- Violation of Title 18 of the United States Code concerning sabotage, espionage, treason, sedition or criminal subversion

- Violation of Articles 94, 104, 106, 133 and 134 of the UCMJ

- Conviction by court-martial during current enlistment

- Absent without leave during current enlistment

- Any proceedings that may result in an administrative elimination

- A written recommendation has been sent to the promotion authority to reclassify a soldier for inefficiency or disciplinary reasons.

- Ineligible to reenlist IAW AR 601-280. Soldier regains promotable status the day he or she receives an approved waiver to reenlist solely because they are within 90 days of ETS will not be considered nonpromotable.

- Soldier without appropriate security clearance or favorable security investigation for promotion to the grade and military occupational specialty (MOS)

- For promotion of Sergeant through Sergeant First Class, when soldiers fail to take a SQT due to their own fault.

- Fail to qualify for reenlistment or extension of their current enlistment to meet the service remaining obligation for promotion to SSG. The promotion authority will remove the names from the promotion recommended list of these soldiers once they have been determined to be ineligible to reenlist or extend.

- Pending a bar to reenlist

- Voluntary retirement application has been approved

- A written recommendation has been submitted to remove the soldier from a promotion recommended list

- Punished under Article 15 of the UCMJ, including suspended punishment, except that any summarized proceedings imposed, according to AR 27-10, paragraph 3-16, are excluded and will not result in a nonpromotable status. The soldier will be promotable on the day following completion of the period of correctional custody, suspension, restriction, extra duty and/or forfeiture of pay, whichever occurs later.

  For the purposes of determining nonpromotable status, periods of forfeiture of pay will be determined as follows: Periods of forfeiture are to begin on the date Article 15 punishment is imposed, for Article 15 forfeitures imposed by company grade commanders, seven calendar days is the period of forfeiture; or Article 15 forfeitures imposed by field grade commanders, fifteen calendar for a one month forfeiture, for two months, the period of forfeiture is forty-five calendar days.

- A physical evaluation board (PEB) determines that a soldier is no longer qualified for continued active service.

- Flagged under the provisions of AR 600-8-2. Failure to initiate DA Form 268 does not negate the fact that a soldier is in a nonpromotable status, if a circumstance exists that requires the imposition of a flag.

- When enrolled in the Army alcohol and drug abuse prevention and control program (ADAPCP), a soldier who would have been promoted while in the program, provided otherwise eligible, will be promoted after successful completion of the program. Date of rank and effective date will be the date the soldier would have been promoted had he or she not been in the program.

DELAY OF PROMOTION DUE TO DA FORM 268

When a delay of promotion has occurred because of suspension of favorable personnel actions, the soldier's promotion status will be determined as follows:

- If the soldier's final report is closed "favorable" and he or she would have been promoted while the suspension of favorable personnel actions was in effect, provided otherwise eligible, the soldier will be promoted. The effective date and date of rank will be that of his or her peers.

- If the soldier's final report is closed "other" (weight and PT) and the soldier would have been promoted while the suspension of favorable action was in effect, provided otherwise eligible, he or she will be promoted with a DOR and effective date of the removal of the suspension of favorable personnel actions.

- Effective date and date of rank stated on a promotion instrument (DA Form 4187) will be the same. The effective date reflected on the promotion instrument will be the effective date used on the grade change transaction.

NCOs SECURITY CLEARANCE REQUIREMENT

The following security clearance requirements are a prerequisite for promotion:

- Promotion to MSG and SGM requires a favorable national agency check (NAC) or a security clearance of secret or higher.

- Promotion to SFC requires the clearance for the promotion MOS

- Promotion to SPC through SSG requires the clearance required by the promotion MOS or an interim clearance at the same level

MOST USABLE TERMS

- **Active Duty**. Full-time duty in the active military service of the United States. It includes full-time training duty, annual training duty and attendance, while in the active military service, at a school designed as a service school by law or by the secretary of the military department concerned. It does not include full-time National Guard duty.

- **Appellate Authority**. Commanders who have final authority to act on appeals.

- **Basic Enlisted Service Date**. Date that reflects total periods of enlisted service, active or inactive, as a member of regular and reserve components of the armed forces of the United States. (Required for computation of enlisted service for promotion to SFC, MSG and SGM)

- **Best Qualified**. Soldiers whom the DA selection board determined to be the best qualified among peers. Also has demonstrated integrity and high moral standards.

- **Commander**. A head of an Army staff or field operating agency or an officer with a position title "Commander" or "Commandant".

- **Command and Staff**. A staff section header by the Commander's senior personnel manager.

- **Creditable Service**. All active or reserve active status service in the grade in which ordered to active duty or higher that may be used to establish DORs under AR 600-8-19.

- **Date of Rank**. The date on which an enlisted soldier was appointed or promoted in a particular grade and the date used to determine relative seniority for soldiers holding the same grade.

- **De Facto Status**. Member who was promoted by competent authority, performed duties of the higher grade, and accepted pay and allowances of the higher grade in good faith and without intent to defraud.

- **Field Promotion Authority**. A commander who may promote enlisted soldiers to the grade PV2-SSG.

- **Grade.** A step or degree, in a graduated scale of office or military rank, that is established and designed as a grade by law or regulation.

- **Military Personnel.** The component of personnel service support that provides that service to soldiers and commanders in the field.

- **Official Military Personnel File.** The official personnel service file usually maintained on microfiche, composed of a performance section, service section and, in some cases, a restricted access section.

- **Policy.** General statement governing objectives of a functional area (within the purview of the office of the Deputy Chief of Staff for Personnel Policy Support).

- **Posthumous Promotion.** A casualty promoted to a higher grade following his or her death.

- **Promotion List.** A list of enlisted soldiers, by grade, recommended and approved for promotion.

- **Promotion Review Authority.** The commander having general court-martial jurisdiction or the first Army General Officer in the chain of command who has a Judge Advocate available.

- **Promotion Sequence Number.** A number that shows the rank order of a soldier on a promotion list.

- **Promotion/Advancement Instrument.** Orders or a DA Form 4187.

- **Rank.** The order of precedence among members of the armed forces.

- **Separation.** Discharge, release from active duty or retirement.

- **Subfunction.** The subdivision of work within functions; for example, the function of enlisted promotions subdivides into enlisted advancements, the semicentralized system and the centralized system.

- **Task**. The major subdivision of a function of subfunction. The lowest level of work which has meaning to the doer. This subdivision has a beginning., an ending and can be measured.

- **Work Center**. Clearly defined organization element recognized by MS3 as the basic for manpower requirements.

APPENDIX B

ABBREVIATIONS

AAM	Army Achievement Medal
ABCMR	Army Board for Correction of Military Records
ACASP	Army Civilian Acquired Skills Program
ACE	American Counsel on Education
ACT	American College Test
ADT	Active Duty for Training
AER	Academic Evaluation Report
AIT	Advanced Individual Training
AMOS	Additionally Awarded Military Occupational Specialty
APFT	Army Physical Fitness Test
ARC	Army Recruiter Course
ARCOM	Army Commendation Medal
ASEP	Advanced Skills Education Program
ASI	Additional Skill Identifier
AWOL	Absent Without Leave
BASD	Basic Active Service Date
BNCOC	Basic Noncommissioned Officer's Course
BNS1	Battalion S-1
BNS2	Battalion S-2
BSEP	Basic Skills Education Program
CDR	Commander
CE	Course Examination
CLEP	College Level Examination Program
CMF	Career Management Field
CPL	Corporal
CPMOS	Career Progression Military Occupational Specialty
CSM	Command Sergeant Major
DA	Department of the Army

DCSS	Declination of Continued Service Statement
DOR	Date of Rank
DCSPER	Deputy Chief of Staff for Personnel
DOD	Department of Defense
EB	Enlistment Bonus
EER	Enlisted Evaluation Report
EMF	Enlisted Master File
EPP	Enlisted Promotion Program
ESL	English as a Second Language
ETS	Expiration Term of Service
FAO	Finance and Accounting Office
FLG	Flagged Records
GCM	General Court-Martial
GCMCA	General Court-Martial Convening Authority
GED	General Education Development
GRCH	Grade Change
HQ	Headquarters
HQDA	Headquarters, Department of the Army
ISR	Individual Soldier's Report
JACT	JUMPS Army Corrector Transaction
LES	Leave and Earning Statement
LTC	Lieutenant Colonel
MACOM	Major Army Command
MCM	Manual for Court-Martial
MIA	Missing in Action
MILPO	Military Personnel Office(s)
MOI	Memorandum of Instruction
MSG	Master Sergeant
MSM	Meritorious Service Medal
MOS	Military Occupational Specialty
MPRJ	Military Personnel Records Jacket
MMPF	Master Military Pay File
NAC	National Agency Check
NCO	Noncommissioned Officer
NCOA	Noncommissioned Officer Academy
OCS	Officer Candidate School
OIC	Officer in Charge
OJE	On-the-Job Experience
OJT	On-the-Job Training
OMPF	Official Military Personnel File
PAB	Personnel Actions Branch

PAC	Personnel and Administration Center
PCS	Permanent Change of Station
PEB	Physical Evaluation Board
PEBD	Pay Entry Basic Date
PED	Promotion Eligibility Date
PERSCOM	U. S. Total Army Personnel Command
P-Fiche	Performance Microfiche
PLC	Primary Leadership Course
PLDC	Primary Leadership Development Course
PMOS	Primary Military Occupational Specialty
PMOSC	Primary Military Occupational Specialty Code
PRMOS	Promotion Military Occupation Specialty
PS	Personnel Support
PSC	Personnel Service Company
PSNCO	Personnel Staff Noncommissioned Officer
QMP	Qualitative Management Program
SA	Secretary of the Army
SFC	Sergeant First Class
SGM	Sergeant Major
SGT	Sergeant
SIDPERS	Standard Installation/Division Personnel System
SMOS	Secondary Military Occupational Specialty
SPC	Specialist
SQI	Skill Qualification Identifiers
SQT	Skill Qualification Test
SRB	Special Reenlistment Bonus
SSG	Staff Sergeant
SSN	Social Security Number
STAB	Standby Advisory Board
TDY	Temporary Duty
TIMIG	Time in Grade
TIS	Time in Service
TJAG	The Judge Advocate General
TRADOC	U. S. Army Training and Doctrine Command
TSO	Test Site Officer
UCMJ	Uniform Code of Military Justice
USAEREC	U. S. Army Enlisted Records and Evaluation Center
USAFAC	U. S. Army Finance and Accounting Center
USAR	U. S. Army Reserve
USMA	United States Military Academy
USMAPS	U. S. Military Academy Preparatory School

INDEX

A
Advisory Board, 89-92
Age, 12
Appeals, 103-104
Appearance, 11-12
Appreciation, 27
Assessment, 61-62
Attitudes:
 changing, 30
 soldiers', 30
Authority, 2
Awards, 79, 81-82

B
Battlefield, 46
Beliefs, 8-9
Buddy team, 41

C
Candor, 6, 32
Ceremonies, 48
Character building, 16
Commissioned Officers, 1
Commitment, 6-7, 33
Communication:
 bottom-up, 61
 competencies, 12
 nonverbal, 11
 verbal, 11
Competence, 6, 33
Constitution, 65-66
Correspondence, 86-87
Counseling, 13
Courage, 6, 33
Criticism, 28

D
Decision making, 14
Delegating, 21
Desire, 34
Doctrine, 64-65
Dress, 78-81

E
Education, 105-118
Ethical:
 dilemmas, 17
 responsibilities, 17
Ethics:
 professional, 15
 situational, 18
Execution, 61-62
Expectations, 42

F
Facial expressions, 12
Fear, 44-45, 50
Feelings, 50
First Sergeant, 19

G
Gender, 12

H
Handshake, 12

I
Inefficiency, 101-102
Initiative, 35
Integrity, 35

J
Judgment, 35
Judgments, 30

K
Knowledge:
   beliefs, 8-9
   self, 3-5
   values, 5-8

L
Leader, 11
Leaders:
   junior, 19
   making, 1-24
Leadership:
   assessing, 4-5
   characteristics of, 4
   competencies, 12-15
   delegating, 21
   directing, 20
   factors of, 10-12
   meaning of, 10
   participating, 21-22
   relationships, 22-24
   skills, 3-4
   styles, 20-22
Led, 10-11
Loyalty, 17, 35

M
Maintenance, 47
Maturity:
   emotional, 34
   physical, 34
   social, 34
   spiritual, 34-35
Meetings with soldiers, 32
METL development, 55-56
Misconduct, 102
Mission, 10, 42, 57-59
Motivation, 10

N
Nameplates, 79
National security, 66
NCO-ER, 94
Noncommissioned Officers, 1-3
Nuclear warfare, 66-67

P
Panic, 51
Personal problems, 38
Planning:
   competency, 15
   informal, 62
   near-term, 60-61
   short-range, 59-60
Platoon sergeants, 10
Preboard, 89
Proficiency:
   tactical, 14
   technical, 14
Promotion:
   centralized, 84-85
   eligibility for, 73-74
   junior NCOs, 71-83
   monthly, 87
   packet, 73
   recommendation for, 73-74
   semicentralized, 71-72
   senior NCO, 84-96
Promotion board:
   centralized, 86
   conducting, 75-76
   study guide for, 76-78
Punishment, 8

R
Reception, 37
Reclassification, 88
Records, 92-93
Reductions, 96-104
Removal, 96-104
Responsibilities, 17
Review, 62-63
Rewards, 39
Ribbons, 79, 81
Rumors, 50
Running, 9

S
Salute, 12
Schools, 105-118
Security clearance, 122
Self-discipline, 35
Senior NCO, 84-96
Sergeant, 83
Sergeant Major Academy, 115-118
Service:
   obligation, 88
   selfless, 35
Simulators, 19-20
Situation, 11
Skin color, 11-12
Soldier teams, 13
Soldiers:
   attitudes of, 30
   caring for, 3, 34, 38
   defining, 26
   expectations of, 4, 46

    fears of, 50
    feelings of, 50
    importance of, 25
    influencing, 26-30
    knowing your, 30-32
    training, 3
    values of, 32-34
Spirit, 34
Squad leader, 40
Standards, 18, 39
Stress, 34
Supervision, 13
Systems, 15

T
Teaching, 13
Team:
    combat-ready, 34-52
    spirit, 47
    small, 43
    training, 53-63
Time in grade, 74-75
Time in service, 74-75
Touching, 12
Trainers, 55
Training:
    combat-level, 54
    computer, 19-20
    evaluating, 62-63
    plans, 58
    principles of, 53-55
    short-range, 59-60
    soldiers, 25-52
Trust, 32
Truthfulness, 32

U
Uniforms:
    Army Green Classic, 78
    men's, 79-81
    women's, 78-81

V
Values
    knowing your, 5-8
    moral, 16

W
War, 65
Warrant Officers, 3
Women, 78-81

ABOUT THE AUTHOR

Master Sergeant (Ret.) Wilson L. Walker began his military writing on active duty in 1982, while stationed in West Germany, when he wrote a study guide for his battalion. His first published book, ***Up or Out: How to Get Promoted as the Army Draws Down***, was primarily written for privates to sergeants (E-5). Not only did the book inform the soldiers as to what needed to be done to get promoted, it also revealed how to go about doing what had to be done.

His next book, ***The Complete Guide to the NCO-ER***, explained why all NCOs should get an excellent report and how to keep their rater and senior rater from giving them a bad report. *"The key to an excellent report is the quarterly counseling,"* says Walker.

When self-development became a big issue, Walker wrote the ***Self-Development Test Study Guide***, which was a combination of the manuals the soldiers had to study for the test. He put the book into a question and answer format with questions and answers from cover to cover of each book.

Walker served in the Army for over 21 years, with a break in service after leaving Vietnam. He served as a section chief, platoon sergeant and first sergeant. Most of his time in service was spent in West Germany (15 years) and tours at Fort Bragg and Fort Bliss. Walker's main concern was always his soldier. *"Give them what they want,"* he said, *"and you will get what you want."* This book represents his many years of experience and philosophy in always putting the soldier first.

CAREER RESOURCES

Contact Impact Publications for a free annotated listing of career resources or visit their World Wide Web site for a complete listing of career resources: *http://www.impactpublications.com*.

The following career resources are available directly from Impact Publications. Complete the following form or list the titles, include postage (see formula at the end), enclose payment, and send your order to:

IMPACT PUBLICATIONS
9104-N Manassas Drive
Manassas Park, VA 20111-2366
Tel. 703/361-7300 or Fax 703/335-9486
E-mail address: impactp@impactpublications.com

Orders from individuals must be prepaid by check, moneyorder, Visa, MasterCard, or American Express. We accept telephone and fax orders.

Qty.	TITLES	Price	TOTAL
Military			
___	Becoming a Better Leader and Getting Promoted in Today's Army	13.95	_____
___	Beyond the Uniform	14.95	_____
___	Complete Guide to the NCO-ER	13.95	_____
___	**CORPORATE GRAY SERIES**	**51.95**	_____
___	■ From Air Force Blue to Corporate Gray	17.95	_____
___	■ From Army Green to Corporate Gray	17.95	_____
___	■ From Navy Blue to Corporate Gray	17.95	_____
___	Guide to Civilian Jobs For Enlisted Naval Personnel	14.95	_____
___	Job Search: Marketing Your Military Experience	16.95	_____
___	Jobs and the Military Spouse	14.95	_____
___	New Relocating Spouse's Guide/Employment	14.95	_____
___	Out of Uniform	12.95	_____
___	Retiring From the Military	25.95	_____
___	Today's Military Wife	16.95	_____
___	Up or Out: How to Get Promoted/Army Draws Down	13.95	_____

Key Directories/Reference Works

	Title	Price	
___	500 Largest U.S. Corporations	14.95	___
___	American Almanac of Jobs and Salaries	20.00	___
___	American Salaries & Wages Survey	105.00	___
___	Big Book of Minority Opportunities	39.95	___
___	Big Book of Opportunities For Women	39.95	___
___	Business Phone Book USA 1997	135.00	___
___	Careers Encyclopedia	39.95	___
___	Complete Directory For People With Disabilities	149.95	___
___	*Complete* Guide For Occupational Exploration	39.95	___
___	Consultants & Consulting Organizations Directory	545.00	___
___	Dictionary of Occupational Titles	39.95	___
___	Directory of Executive Recruiters 1997	44.95	___
___	Directory of Federal Jobs and Employers	21.95	___
___	Encyclopedia of Associations 1997	1,149.00	___
___	Encyclopedia of Careers/Vocational Guidance	149.95	___
___	*Enhanced* Guide For Occupational Exploration	34.95	___
___	Government Phone Book USA 1997	185.00	___
___	Guide to Internet Databases	114.00	___
___	**HOOVER'S KEY EMPLOYER DIRECTORIES**	**141.95**	___
___	▪ Hoover's 500	29.95	___
___	▪ Hoover's Emerging Companies 1996	29.95	___
___	▪ Hoover's Guide to Computer Companies	34.95	___
___	▪ Hoover's Handbook of World Business	27.95	___
___	▪ Hoover's Top 2,500 Employers	22.95	___
___	Internships 1997	24.95	___
___	**JOB FINDERS FOR 1997**	**50.95**	___
___	▪ Government Job Finder	16.95	___
___	▪ Nonprofit's and Education Job Finder	16.95	___
___	▪ Professional's Job Finder	18.95	___
___	Job Hunter's Sourcebook	69.95	___
___	Job Hunter's Yellow Pages	35.00	___
___	Jobs Rated Almanac	16.95	___
___	Moving & Relocation Sourcebook	179.95	___
___	National Job Hotline Directory 1997	14.95	___
___	National Trade & Professional Associations	85.00	___
___	Occupational Outlook Handbook	16.95	___
___	Personnel Executives Contactbook	149.00	___
___	Professional Careers Sourcebook	99.95	___
___	Training & Development Organizations Directory	389.00	___
___	U.S. Industrial Outlook	29.95	___
___	Vocational Careers Sourcebook	84.95	___

City and State Job Banks

	Title	Price	
___	**METROPOLITAN EMPLOYER CONTACT DIRECTORIES KIT** (44 titles)	**764.95**	___
___	▪ Atlanta (Job Bank)	16.95	___
___	▪ Atlanta (How to Get a Job in)	16.95	___
___	▪ Austin/San Antonio (Job Bank)	16.95	___
___	▪ Boston (Job Bank)	16.95	___
___	▪ Carolina (Job Bank)	15.95	___

CAREER RESOURCES   135

	▪ Cincinnati (Job Bank)	16.95	____
___	▪ Chicago (Job Bank)	16.95	____
___	▪ Chicago (How to Get a Job in)	16.95	____
___	▪ Chicago Area Companies (Hoover's Guide...)	24.95	____
___	▪ Cleveland (Job Bank)	16.95	____
___	▪ Dallas/Fort Worth (Job Bank)	16.95	____
___	▪ Denver (Job Bank)	15.95	____
___	▪ Detroit (Job Bank)	16.95	____
___	▪ Europe (How to Get a Job in)	17.95	____
___	▪ Florida (Job Bank)	16.95	____
___	▪ Houston (Job Bank)	16.95	____
___	▪ Indianapolis (Job Bank)	16.95	____
___	▪ Las Vegas (Job Bank)	16.95	____
___	▪ Los Angeles (Job Bank)	16.95	____
___	▪ Minneapolis/St. Paul (Job Bank)	16.95	____
___	▪ Missouri (Job Bank)	16.95	____
___	▪ New Mexico (Job Bank)	16.95	____
___	▪ New York (Job Bank)	16.95	____
___	▪ New York (How to Get a Job in)	16.95	____
___	▪ New York Area Companies (Hoover's Guide...)	24.95	____
___	▪ North New England (Job Bank)	16.95	____
___	▪ Ohio (Job Bank)	16.95	____
___	▪ Philadelphia (Job Bank)	16.95	____
___	▪ Phoenix (Job Bank)	15.95	____
___	▪ Pittsburgh (Job Bank)	16.95	____
___	▪ Portland (Job Bank)	16.95	____
___	▪ San Francisco (Job Bank)	16.95	____
___	▪ San Francisco (How to Get a Job in)	16.95	____
___	▪ Seattle (Job Bank)	16.95	____
___	▪ Seattle/Portland (How to Get a Job in)	16.95	____
___	▪ Southern California (How to Get a Job in)	16.95	____
___	▪ Southern California Area Companies (Hoover's...)	24.95	____
___	▪ Tennessee (Job Bank)	16.95	____
___	▪ Texas Area Companies (Hoover's Guide...)	24.95	____
___	▪ Upstate New York (Job Bank)	16.95	____
___	▪ Virginia (Job Bank)	16.95	____
___	▪ Washington, DC (Job Bank)	16.95	____

Using the Internet and Computers

___	Be Your Own Headhunter Online	16.00	____
___	Electronic Job Search Revolution	12.95	____
___	Electronic Resume Revolution	12.95	____
___	Electronic Resumes: Putting Your Resume On-Line	19.95	____
___	Electronic Resumes For the New Job Market	11.95	____
___	Finding a Job On the Internet	16.95	____
___	Hook Up, Get Hired	12.95	____
___	How to Get Your Dream Job Using the Internet	29.99	____
___	Net Jobs: How to Use the Internet	22.00	____
___	Point and Click Jobfinder	14.95	____
___	Selling On the Internet	24.95	____
___	Three-Rs of E-Mail	12.95	____
___	Using the Internet and the WWW in Your Job Search	16.95	____

Finding Great Jobs and Careers

____	100 Best Careers For the 21st Century	15.95	_____
____	100 Fastest Growing Companies in America	14.95	_____
____	101 Great Answers/Toughest Job Search Problems	11.99	_____
____	101 Ways to Power Up Your Job Search	12.95	_____
____	110 Biggest Mistakes Job Hunters Make	15.95	_____
____	303 Off the Wall Ways to Get a Job	12.99	_____
____	Adams Jobs Almanac 1997	15.95	_____
____	Adventure Careers	11.99	_____
____	American Almanac of Jobs & Salaries	20.00	_____
____	America's Top Jobs Book Plus CD-ROM	39.95	_____
____	Best Jobs For the 1990s & Into the 21st Century	19.95	_____
____	But What If I Don't Want to Go to College	10.95	_____
____	Career Atlas	12.99	_____
____	Careers in Computers	17.95	_____
____	Careers in Education	17.95	_____
____	Careers in Health Care	17.95	_____
____	Careers in High Tech	17.95	_____
____	Careers in Multimedia	24.95	_____
____	Change Your Job, Change Your Life	17.95	_____
____	Complete Idiot's Guide to Getting the Job You Want	24.95	_____
____	Complete Job Finder's Guide to the 90's	13.95	_____
____	Dare to Change Your Job and Your Life	14.95	_____
____	Directory of Executive Recruiters 1997	44.95	_____
____	Five Secrets to Finding a Job	12.95	_____
____	Free and Inexpensive Career Materials	19.95	_____
____	Get a Job You Love!	19.95	_____
____	Hidden Job Market 1997	18.95	_____
____	Hoover's Top 2,500 Employers	22.95	_____
____	How to Get Interviews From Classified Job Ads	14.95	_____
____	How to Make Use of a Useless Degree	13.00	_____
____	How to Succeed Without a Career Path	13.95	_____
____	In Transition	12.50	_____
____	Job Finding Skills For Smart Dummies	37.95	_____
____	Job Hunter's Word Finder	12.95	_____
____	Job Hunting For Dummies	16.99	_____
____	Job Hunter's Catalog	10.95	_____
____	Jobs 1997	16.00	_____
____	Jobs and Careers With Nonprofit Organizations	15.95	_____
____	Jobs For Lawyers	14.95	_____
____	Jobs Rated Almanac	16.95	_____
____	Joyce Lain Kennedy's Career Book	29.95	_____
____	Knock 'Em Dead 1997	12.95	_____
____	_New_ Complete Guide to Environmental Careers	15.95	_____
____	_New_ Relocating Spouse's Guide to Employment	14.95	_____
____	Nonprofits and Education Job Finder	16.95	_____
____	Part-Time Jobs	32.95	_____
____	Professional's Job Finder	18.95	_____
____	Rites of Passage at $100,000+	29.95	_____
____	Sunshine Careers	16.95	_____
____	Top 10 Fears of Job Seekers	12.00	_____
____	Very Quick Job Search	14.95	_____

CAREER RESOURCES   137

___ What Color Is Your Parachute? 1997   16.95   _____
___ World Almanac Job Finder's Guide 1997   24.95   _____

Cover Letters

___ 175 High-Impact Cover Letters   10.95   _____
___ 200 Letters for Job Hunters   19.95   _____
___ 201 Dynamite Job Search Letters   19.95   _____
___ 201 Killer Cover Letters   16.95   _____
___ 201 Winning Cover Letters for $100,000+ Jobs   24.95   _____
___ Adams Cover Letter Almanac and Disk   19.95   _____
___ Cover Letters For Dummies   12.99   _____
___ Cover Letters That Knock 'Em Dead   10.95   _____
___ Dynamite Cover Letters   14.95   _____

Resumes

___ 100 Winning Résumés for $100,000+ Jobs   24.95   _____
___ 101 Great Résumés   9.99   _____
___ 175 High-Impact Résumés   10.95   _____
___ Adams Résumé Almanac   10.95   _____
___ Asher's Bible of Executive Résumés   29.95   _____
___ Best Résumés for $75,000+ Executive Jobs   14.95   _____
___ Complete Idiot's Guide to Crafting the Perfect Résumé 16.95 _____
___ Designing the Perfect Résumé   12.95   _____
___ Dynamite Résumés   14.95   _____
___ Electronic Résumé Revolution   12.95   _____
___ Electronic Résumés: Putting Your Résumé On-Line   19.95   _____
___ Electronic Résumés for the New Job Market   11.95   _____
___ Encyclopedia of Job-Winning Résumés   16.95   _____
___ Gallery of Best Résumés   16.95   _____
___ Gallery of Best Résumés for Two-Year Degree Graduates 14.95 _____
___ High Impact Résumés and Letters   14.95   _____
___ Résumé Catalog   15.95   _____
___ Résumé Kit   9.95   _____
___ Résumé Shortcuts   14.95   _____
___ Résumé Solution   12.95   _____
___ Résumés for Advertising Careers   9.95   _____
___ Résumés for Architecture and Related Careers   9.95   _____
___ Résumés for Banking and Financial Careers   9.95   _____
___ Résumés for Business Management Careers   9.95   _____
___ Résumés for Communications Careers   9.95   _____
___ Résumés for Dummies   12.99   _____
___ Résumés for Education Careers   9.95   _____
___ Résumés for Engineering Careers   9.95   _____
___ Résumés for Environmental Careers   9.95   _____
___ Résumés for Ex-Military Personnel   9.95   _____
___ Résumés for 50+ Job Hunters   9.95   _____
___ Résumés for First-Time Job Hunter   9.95   _____
___ Résumés for the Healthcare Professional   12.95   _____
___ Résumés for High Tech Careers   9.95   _____
___ Résumés for Midcareer Job Changers   9.95   _____
___ Résumés for the Over 50 Job Hunter   14.95   _____

___ Résumés for Re-Entering the Job Market   9.95   _____
___ Résumés for Sales and Marketing Careers   9.95   _____
___ Résumés for Scientific and Technical Careers   9.95   _____
___ Résumés That Knock 'Em Dead   10.95   _____

Skills, Testing, Self-Assessment, Empowerment

___ 7 Habits of Highly Effective People   14.00   _____
___ Discover the Best Jobs for You   11.95   _____
___ Do What You Are   14.95   _____
___ Do What You Love, the Money Will Follow   10.95   _____
___ Love Your Work and Success Will Follow   12.95   _____

Dress and Etiquette

___ 110 Mistakes Working Women Make...   9.95   _____
___ Dress Casually For Success .... For Men   16.95   _____
___ Executive Etiquette in the New Workplace   14.95   _____
___ John Molloy's New Dress For Success (Men)   13.99   _____
___ *New* Women's Dress For Success   12.99   _____
___ Red Socks Don't Work!   14.95   _____
___ Winning Image   17.95   _____

Networking and Power Building

___ Dynamite Networking For Dynamite Jobs   15.95   _____
___ Dynamite Tele-Search   12.95   _____
___ Great Connections   19.95   _____
___ How to Work a Room   11.99   _____
___ NBEW's Networking   10.95   _____
___ Power Networking   14.95   _____
___ Power Schmoozing   12.95   _____
___ Power to Get In   24.95   _____
___ Secrets of Savvy Networking   12.99   _____

Interviewing

___ 50 Winning Answers to Interview Questions   10.95   _____
___ 60 Seconds and You're Hired   9.95   _____
___ 90-Minute Interview Prep Book   15.95   _____
___ 101 Dynamite Questions to Ask at Your Job Interview 14.95 _____
___ 101 Great Answers/Interview Questions   9.99   _____
___ 111 Dynamite Ways to Ace Your Job Interview   13.95   _____
___ Adams Job Interview Almanac   10.95   _____
___ Best Answers to 201 Most/Asked Interview Questions 10.95 _____
___ Conquer Interview Objections   10.95   _____
___ Dynamite Answers to Interview Questions   11.95   _____
___ Dynamite Salary Negotiations   13.95   _____
___ Interview For Success   15.95   _____
___ Interview Kit   10.95   _____
___ Interview Power   12.95   _____
___ Job Interviews For Dummies   12.99   _____

___ Killer Interviews	10.95	_____
___ Naked At the Interview	10.95	_____
___ NBEW's Interviewing	11.95	_____
___ Perfect Follow-Up Method to Win the Job	12.95	_____
___ Power Interviews	12.95	_____
___ Quick Interview and Salary Negotiation Book	12.95	_____

SUBTOTAL _____

Virginia residents add 4½% sales tax _____

POSTAGE/HANDLING ($5.00 for first title and $1.50 for each additional book) __$5.00__

Number of additional titles x $1.50----------------- _____

TOTAL ENCLOSED -------------------- _____

SHIP TO:

NAME _____

ADDRESS _____

_____

❑ I enclose check/moneyorder for $ _____ made payable to IMPACT PUBLICATIONS.

❑ Please charge $ _____ to my credit card:

   ❑ Visa   ❑ MasterCard   ❑ American Express

   Card # _____

   Expiration date: _____ / _____

   Signature _____

We accept official purchase orders from libraries, educational institutions, and government offices. Please attach copy with official signature(s).

The On-Line Superstore & Warehouse
Hundreds of Terrific Career Resources Conveniently Available On the World Wide Web 24-Hours a Day, 365 Days a Year!

Ever wanted to know what are the newest and best books, directories, newsletters, wall charts, training programs, videos, CD-ROMs, computer software, and kits available to help you land a job, negotiate a higher salary, or start your own business? What about finding a job in Asia or relocating to San Francisco? Are you curious about how to find a job 24-hours a day by using the Internet or what to do after you leave the military? Trying to keep up-to-date on the latest career resources but not able to find the latest catalogs, brochures, or newsletters on today's "best of the best" resources?

Welcome to the first virtual career bookstore on the Internet. Now you're only a "click" away with Impact Publication's electronic solution to the resource challenge. Impact Publications, one of the nation's leading publishers and distributors of career resources, has launched its comprehensive "Career Superstore and Warehouse" on the Internet. The bookstore is jam-packed with the latest resources focusing on several key career areas:

- Alternative jobs and careers
- Self-assessment
- Career planning and job search
- Employers
- Relocation and cities
- Resumes
- Cover Letters
- Dress, image, and etiquette
- Education
- Telephone
- Military
- Salaries
- Interviewing
- Nonprofits
- Empowerment
- Self-esteem
- Goal setting
- Executive recruiters
- Entrepreneurship
- Government
- Networking
- Electronic job search
- International jobs
- Travel
- Law
- Training and presentations
- Minorities
- Physically challenged

The bookstore also includes a new "Military Career Transition Center" for the latest in resources relevant to all service members and their families.

"This is more than just a bookstore offering lots of product," say Drs. Ron and Caryl Krannich, two of the nation's leading career experts and authors and developers of this on-line bookstore. *"We're an important resource center for libraries, corporations, government, educators, trainers, and career counselors who are constantly defining and redefining this dynamic field. Of the thousands of career resources we review each year, we only select the 'best of the best.'"*

Visit this rich site and you'll quickly discover just about everything you ever wanted to know about finding jobs, changing careers, and starting your own business—including many useful resources that are difficult to find in local bookstores and libraries. The site also includes what's new and hot, tips for job search success, and monthly specials. Impact's Web address is:

http://www.impactpublications.com